S. S. Rakhimkhodjaev
G. N. Sobirova

THEORY OF FABRIC CONSTRUCTION: Single-layer weaves

S. S. Rakhimkhodjaev
G. N. Sobirova

THEORY OF FABRIC CONSTRUCTION: Single-layer weaves

ScienciaScripts

Imprint

Any brand names and product names mentioned in this book are subject to trademark, brand or patent protection and are trademarks or registered trademarks of their respective holders. The use of brand names, product names, common names, trade names, product descriptions etc. even without a particular marking in this work is in no way to be construed to mean that such names may be regarded as unrestricted in respect of trademark and brand protection legislation and could thus be used by anyone.

Cover image: www.ingimage.com

This book is a translation from the original published under ISBN 978-620-7-48663-2.

Publisher:
Sciencia Scripts
is a trademark of
Dodo Books Indian Ocean Ltd. and OmniScriptum S.R.L publishing group

120 High Road, East Finchley, London, N2 9ED, United Kingdom
Str. Armeneasca 28/1, office 1, Chisinau MD-2012, Republic of Moldova, Europe
Managing Directors: Ieva Konstantinova, Victoria Ursu
info@omniscriptum.com

Printed at: see last page
ISBN: 978-620-8-53120-1

Contents

OUTLINE.

The paper presents the theory of structure and peculiarities of dressing production of fabrics of main, derived from main and combined weaves. Definitions, purposes, advantages, disadvantages, principles, types of these weaves are given. The methods of construction of the main, derived from the main and combined weaves are given. The principles of construction of these weaves are substantiated. Examples of construction and analysis of fabrics of main, derived from main and combined weaves are shown. It is intended for researchers, technologists, designers, constructors, dessinators, masters and bachelors, textile industry workers engaged in the study of main, derived from main and combined weaves.

INTRODUCTION

Weaving as an art and craft has deep roots. In distant times, man created various objects for the convenience of existence - clothes, shoes, bedding, baskets, nets, etc. Obtaining these objects (products) was carried out by weaving strips of animal skin, grass, reeds, flax, bushes and trees. Such creativity of ancestors created weaving - one of the forms of weaving and weaving devices - weaving frame, hand looms with vertical, horizontal and circular arrangement of warp. The design of looms was determined by the type of material to be processed, the weave of the fabric, climatic conditions and the way of life of the people. Nowadays a modern weaving machine is high-speed, computerised, with excellent ergonomics, with a wide assortment of fabrics produced with operative change of assortment, with production of high quality fabrics. Almost all peoples of the world have myths and legends connected with fabric manufacturing, which were reflected in the literature and art of the time. An example of this is a verse treatment of the ancient Iranian story "Shahnameh" by Ferdowsi - for the dressing of silk, furs, cloth, from cocoons, skins and light linen, spinning threads taught him and standing behind the machine, to weave cleverly into the base of the duck Man has always strived to make his clothes dressy and comfortable. Pattern on the fabric can be obtained by weaving by interweaving warp and weft threads and in the process of finishing the finished fabric, embroidery or printing. Pattern formation by weaving is accompanied by artistic and technological process, the combination of which allows to achieve a variety of light and texture effects in the pattern, through different reflection of light from different parts of the fabric. A distinction is made between single-layer and multi-layer fabrics. Single-layer fabrics are produced only on the basis of main weaves, derived from main weaves, combined weaves and simple jacquard weaves. Multilayer fabrics are produced only on the basis of complex weaves and complex jacquard weaves. Main weaves, derived from main weaves, combined weaves and simple jacquard weaves are considered simple because they are constructed with one main thread system and one weft thread system. Complex weaves are those with several systems of main threads and several systems of weft threads. Each of the thread systems is placed one above the other, forming layers of fabric. Therefore, the front and back sides of the fabric are independent weaves.

1. TISSUE STRUCTURE AND ANALYSIS

A fabric is a textile product obtained by weaving two systems of threads (at least) arranged mutually perpendicular to each other. Threads located along the length of the fabric are called warp, and those located along the width of the fabric are called weft (Fig. 1). The filling pattern of the fabric is depicted on a chequered (kanvov) paper, where vertical interstitches mean warp threads and horizontal interstitches mean weft threads. (Fig.1).

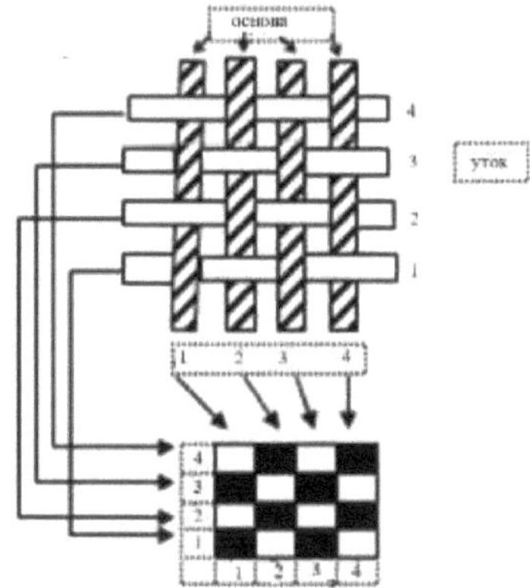

Figure 1. Filling pattern of the fabric.

The places where the warp and weft threads intersect are called overlaps. If the warp is above the weft, the overlap is the main overlap and is indicated by a coloured box. If the warp is under the weft, the overlap is the weft overlap and is indicated by an uncoloured box. The mutual overlapping of yarns of one system with yarns of another system is called the weave of the fabric. The parameters of the weave of a fabric are the weave rapport and the weave shift. By varying the weave shift and weave rapport, the structure of the fabric can be changed. The weave length is the smallest number of threads after which the overlapping starts again (Fig. 2). A distinction is made between the warp rapport R_o and the weft rapport R_y. The warp rapport R_o is the smallest number of warp threads after which the overlapping starts again. The weft ratio R_y is the smallest number of weft yarns after which the overlap begins to repeat. The weave shift is determined by the distance between single overlaps on neighbouring yarns of the same system (Fig. 2). If the distance between single overlaps on neighbouring main yarns is determined, the shift will be on the base s_o. If the distance between single overlaps on neighbouring weft yarns is determined, the shear will be in the weft s_y.

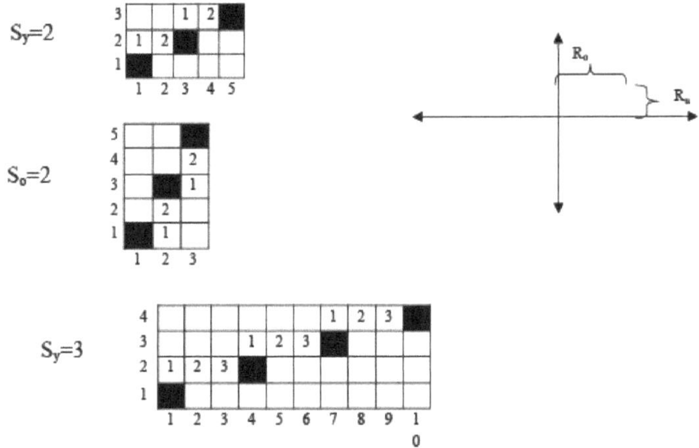

Figure 2. Parameters of the weave of the fabric.

For threading and weaving on the weaving machine, a filling pattern is drawn up. The threading pattern contains the weave pattern, the main threads in the reed, the warp threads, and the warp threading pattern
to the remise, the order of moving the remise (cardboard) and the schemes of fabric cuts on the warp and weft (Fig.3). The construction of the filling pattern starts with the image of the weave pattern of the fabric within one weave rapport (Fig. 4). Depending on the type of weave of the fabric, the type of fringing of the main threads in the remise and the number of remises in the dressing are chosen. Then, a diagram of the reed and remiz threading is shown.

Порядок подъема ремиз, правая каретка	Проборка в ремиз	Порядок подъема ремиз левая каретка
	Проборка в бердо	Разрез ткани вдоль основы
	Рисунок переплетения	

Разрез ткани вдоль утка

Remise lift order, right-hand carriage
Punching in the remise
Order of lifting remise left carriage
Reed furring Weave pattern
Cutting the fabric along the base
Fabric cut along the weft

Fig.3.Full filling pattern of the fabric.

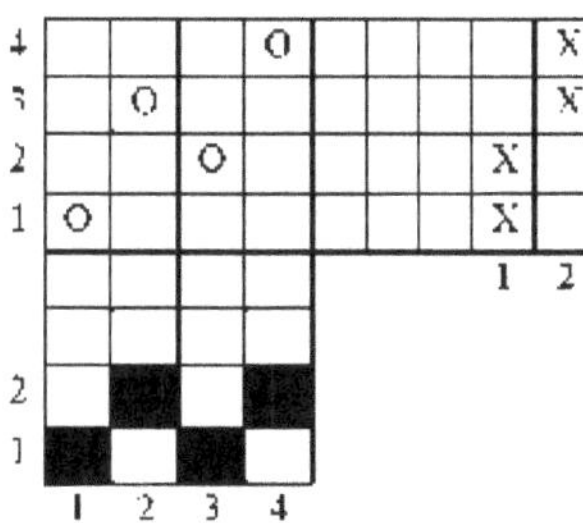

Figure 4. Complete filling pattern of a fabric of a given weave.

After that, the order of the heddles for each weft sashing within the weave pattern is determined and shown. Next, the fabric cuts are made on the warp and weft. When drawing up the filling pattern, the number of heddles in the filling pattern is equal to the number of warp yarns that are intertwined differently, as the warp yarns that are equally intertwined along the entire length of the pattern are picked up in the same heddle. Also, the number of cards in the filling pattern is equal to the number of weft yarns in the weave pattern.

Fabric analysis consists of examining a fabric sample in order to obtain the data required for the construction of the filling pattern and the calculation of the fabric filling. A ruler, weaving magnifying glass, needle, scissors are used as tools. In this case determine: warp and weft yarns, face and underside of the fabric, the density of the fabric on the warp and weft, working of warp and weft yarns, linear density of warp and weft, weave of the fabric.

The classification of fabrics according to the number of yarn systems used and the type of weave is given in the following order. Fabrics produced from two thread systems (warp and weft) are single-layer and the weave of these fabrics can have main, derived, combined and jacquard (large-patterned) simple. Fabrics made of at least three yarn systems (one warp and two wefts or vice versa) are multilayer and the weave of these fabrics can be complex and jacquard (large-patterned) complex.

2. WARP STRIPPING IN THE REMISE

Depending on the type of weave, the number of heddles and the order in which the warp yarns are threaded into the heddles varies.

The number of heddles (k) in a filling depends on the number of differently interlaced yarns in the warp weave ratio (R_0) and the warp density of the fabric (R_0). Therefore, the types of stitches used in the remise are subdivided according to values such as the warp rapport (R_0), the stitch rapport (**r**), and the number of remise (**k**). Depending on the ratio of these values, all types of corks are divided into three groups:

1. R_0 r= = **to**,   2. R_0 r<= **to**,   3. R_0 r= > **to**.

The length of the sashing is the smallest number of warp threads whose sashing order is repeated in a remise. For free movement of the warp and warp threads of one remise relative to another remise, it is necessary to have a permissible density of the warp threads on the remise, which depends on the thickness of the warp threads and the type of warp threads. To the first group of gilets are row gilets where R_0 r= = **k**, e.g. **R0** = 5, **r** = 5, **k** 5.

	1	2	3	4	5	1	2	3	4	5	1	2	3	4	5
5					0					0					0
4				0					0					0	
3			0					0					0		
2		0					0					0			
1	0					0					0				

Figure 1.

In row picking, the warp yarns are picked into the galeva remise one by one, starting with the first remise. Therefore, the first warp is picked up in the first stitch, the second warp in the second stitch, and so on until the picking pattern is completed. The number of purl rapports depends on the number of threads in the warp. The rapport of the row piercing is equal to the number of remisques **r= k**, and the number of remisques is equal to the rapport of the weave in the warp **to= R_0**. In our example (Figure 1), the row purl has R_0 = 5, **k** = 5, **r** 5. The row purl can be used in all cases, i.e. it can produce any weave. <u>The advantage of the </u>row picking is that it is easy and convenient for the weaver to operate when threading and eliminating warp breaks. The disadvantage of row picking is that the increase of the warp weave rapport leads to an increase in the number of remisks in the threading. In addition, with a high warp density, the density of the warp will increase, which leads to increased warp breakage during weaving.

The second group includes a scattering of R_0 r<= **to**.

In this picking, the warp threads are picked through one (two, three, etc.) remizki in the galeva, i.e. in the beginning they are picked in all odd-numbered remizki, and then in all even-numbered remizki, after which the picking is repeated. For example, if the rapport on the base $R_0 = 2$, and the number of remizki $k = 4$, then picking is carried out in the following sequence (Fig. 2): the first thread in the first remizki; the second thread in the third remizki; the third thread in the second remizki; the fourth thread in the fourth remizki.

4				0				0				0
3		0				0				0		
2			0				0				0	
1	0				0				0			
	1	2	3	4	1	2	3	4	1	2	3	4

Figure 2.

Consequently, the length of the striping r is equal to the number of reams k, and the number of reams is equal to twice the value of the warp length $K = 2R_0$. It is expedient to use the scattered picking when producing warp-dense fabrics, as the density of the gules on one remizki decreases, and the rubbing of yarns and gules close to each other is prevented, as a result of which the breakage of warp yarns is reduced.

The third group of selvedges includes reverse selvedges, composite selvedges and reduced selvedges. If there are equally interlaced yarns in the warp pattern R_0, a purl is used with a reduced number of cuts compared to the warp pattern, i.e. $R_0 r = >$ **to**.

In reverse plain picking (Fig. 3), the warp yarns are picked as in row picking (straight diagonal) and then in the opposite direction (reverse diagonal), etc. The number of threads in the reverse plain sashing rapport $r = 2k - 2$, **R0** $= 8$, $r = 8$, $k = 8$. The example of Fig.3 shows a selvedge with $R_0 = 8$, $r = 8$ and the number of remisks **k** $= 5$.

5					0								0			
4				0		0						0		0		
3			0				0				0				0	
2		0						0		0						0
1	0								0							
	1	2	3	4	5	6	7	8	1	2	3	4	5	6	7	8

Figure 3. Reverse simple punching.

In the case of reverse double selvedge, the selvedge pattern is equal to twice the number of remisks, i.e. the number of selvedged main threads is twice as much as the number of remisks (Fig. 4) $r = 2k$. In the example of Fig.4 **R0** $= 12$, $r = 12$, $k = 6$.

6						0	0											0
5					0			0									0	
4				0					0							0		
3			0							0					0			
2		0									0			0				
1	0											0	0					
	1	2	3	4	5	6	7	8	9	10	11	12	1	2	3	4	5	6

Figure 4. Reverse double stripe in remise.

Reverse piercing on the pattern in remiz (Fig. 5) as we see has repeated straight and reverse diagonals in the piercing rapport such number, which correspond to the number of strips of the same weave, so in the rapport of this piercing significantly increases the number of sampled warp threads without increasing the number of remizki. In our example $R_0 = 22$, $r = 22$, $k = 4$.

	1	2	3	4	5	6	7	8	9	10	11	12	13	14	15	16	17	18	19	20	21	22
4				0				0				0				0				0		
3			0				0				0		0				0				0	
2		0				0				0				0				0				0
1	0				0				0						0				0			

Fig. 5. Reverse punching on the pattern in remise.

Reverse striping in remise is used for fabrics with symmetrical longitudinal patterns and weaves.

Rémise to rémise is used when weaving fabrics with two or more different weaves. Rhemisks are divided into vaults (Fig.6 and Fig.7), containing different or equal number of rhemisks. The number of vaults in a dressing is equal to the number of weaves in the fabric. There are continuous (Fig.6) and continuous (Fig.7) vaults of the main threads in a remise.

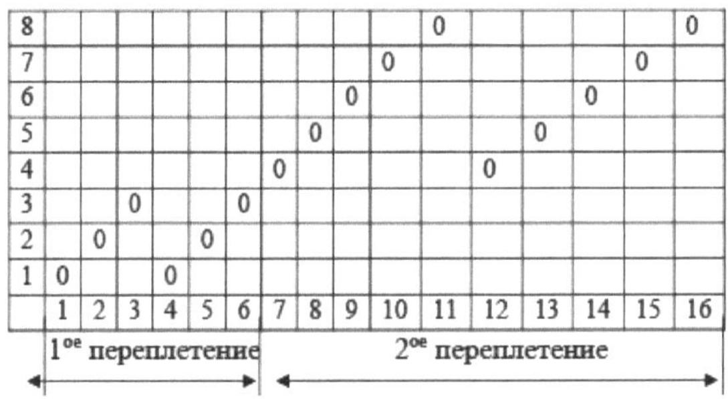

	1	2	3	4	5	6	7	8	9	10	11	12	13	14	15	16
8											0					0
7										0					0	
6									0					0		
5								0					0			
4							0					0				
3			0			0										
2		0			0											
1	0			0												

1st weave 2nd weave

Figure 6. Consolidated discontinuous punching in remise.

In vaulted interrupted picking in remise (Fig. 6), the warp yarns are picked first in one vault and then in another vault. In our example, for the first weave the warp pattern $R_{01} = 6$ and the warp is picked in three heddles (first heddle), and for the second weave the warp pattern $R_{02} = 10$ and the warp is picked in the next five heddles (second heddle). The total weave rapport $R0 = 16$, and the total number of remisks $k = 8$, the sampling rapport r 16. Consolidated discontinuous fringing is used in the production of fabrics with longitudinal or square stripes of various types of weaves, repeated several times in the strip.

In consolidated continuous sashing, the warp yarns of one weave are placed (picked through) between the warp yarns of the other weave (Fig. 7).

8								0								8	0	

	1	2	3	4	5	6	7	8	9	10	11	12	13	14	15	16
7						0								0		
6				0								0				
5		0								0						
4							0								0	
3					0								0			
2			0								0					
1	0								0							

Figure 7. Consolidated continuous probit in remise.

Fig. 7 shows that two types of weaves are used for the fabric production and the warp threads of each weave are selvedged into the vaults by alternating the selvedges of each thread. The basis rapport for the first vault and the second vault for our case $R_{01} = R_{02} = 4$, the number of remisks for each vault is four. The total base rapport $R_0 = R_{01} + R_{02} = 4 + 4 = 8$, the number of reims $k = 8$.

Reduced (by pattern) picking is used when weaving fine-patterned weave fabrics with a large number of warp yarns of the same weave in the rapport, which are picked in the same remizka (Fig.8).

	1	2	3	4	5	6	7	8	1	2	3	4	5	6	7	8
4						0	0							0	0	
3					0			0					0			0
2		0	0							0	0					
1	0			0					0			0				

Figure 8. Abbreviated proborka in remise.

Disadvantages of the third group of corks: unequal number of galleys on the ribs; unequal load on individual ribs; complexity of corking.

unequal number of galleys on the ribs; unequal load on individual ribs; complexity of corking.

3.MAIN (FUNDAMENTAL) WEAVES

A main weave is a weave in which each warp and weft thread overlaps or is overlapped by only one thread of the opposite system within the rapport, i.e. within the rapport there is one main overlap for each thread among the other weft threads or one weft overlap among the other main overlaps. The overlap shift is constant S_o= const, S_y= const and the weave rapports are R_o= R_y. There are three types of main weaves, each of which is defined by its own parameters: plain; twill; satin (satin).

plain weave

The weave is characterised by the fact that each weft thread is interwoven with each warp thread. The parameters of plain weave fabrics are R_o= R_y = 2, S_o= S_y = 1. The plain weave fabrics have the same number of warp and weft overlaps on the front and back sides. The appearance of plain weave fabrics depends on the following factors: warp and weft tension (determines the amount of bending of the yarns of one system relative to the other, i.e. phase of fabric structure); warp and weft densities (a fabric with a high warp density has longitudinal scars on the surface, and a fabric with a high weft density has transverse scars on the surface); the ratio of linear densities of warp (T_o) and weft (T_u), (at T_o> T_u longitudinal scars appear on the surface of the fabric, and at T_u> T_u transverse scars); the size of the backstitch and the position of the scalo in height (with increasing backstitch and changing the position of the scalo in height, the surface of the fabric has a more stiff and even structure); twist direction of warp and weft yarns (at the same twist direction of warp and weft yarns the fabric surface has a more pronounced plain weave pattern); from the values and twist direction in warp or weft yarns the use of two warp yarns with different twist (**Z**) and (**S**) forms a fabric with corrugated surface (and the use of two types of weft yarns with different twist (**Z**) and (**S**) forms a fabric with crepe effect on the surface); from the type and colour of warp and weft threads (the use of smooth shiny threads of different (contrasting) colours in the warp and weft forms a multi-coloured iridescence on the surface of the fabric); the appearance of the finished fabric depends on the method and type of its finishing. The production of plain weave fabric requires only two remizki, but the fabric will have a low density on the basis. Therefore, when producing denser

In plain weave fabrics, four, six, and sometimes eight heddles are used for punching. The fabric is produced using row (Fig. 1) and scattered picking (Fig. 2). The scattered sashing is the most rational, as three side by side sashes are moved simultaneously, which reduces the friction between the sashes and yarns, which reduces the breakage of warp yarns and increases the service life of sashes.

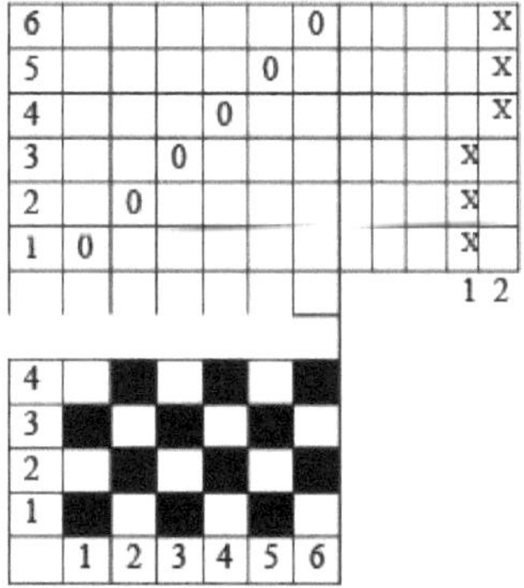

Fig.1.Filling pattern of plain weave.

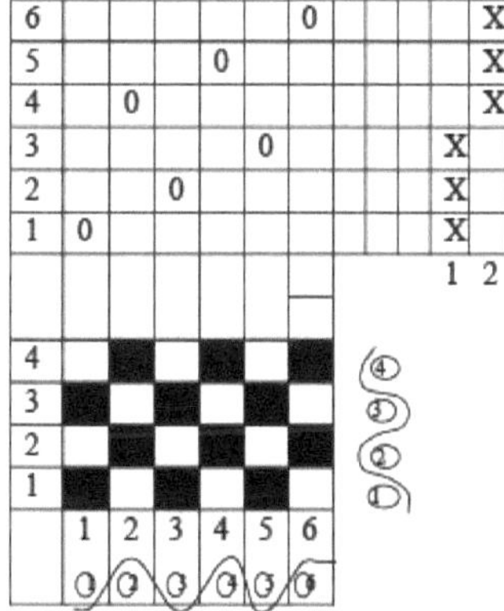

Figure 2. Filling pattern of plain weave.

<h1 style="text-align:center">Twill weave</h1>

A characteristic feature of twill is that the overlap of each main thread is offset, forming inclined diagonal lines (steps) at an angle of 45^0, running from left to right. The smallest rapport of a twill weave is three threads $R_o= R_y> 2$, and the maximum rapport is determined by the size of the planking, the length of which must not exceed 4 mm. twill shift $S_o= S_y= \pm 1$. Twill weave is denoted by a fraction where the numerator shows the number of main overlaps and the denominator the number of weft overlaps, the sum of the numerator and denominator is the rapport of the weave. The front and back sides of the fabric are not the same because the length of the main and weft overlaps are different and the diagonal stripes have different directions. Fig.3, shows the complete filling pattern of twill 1/3 and twill 3/1 (Fig.4). The ratio of the weave is the sum of the numerator and denominator:

$R_o= R = 1 + 3 = 4$, $S_o= S_u= 1$ (Fig.3)

$R_o= R = 3 + 1 = 4$, $S_o= S_u = -1$ (Fig.4)

If weft overlaps predominate on the front side of the fabric, the twill is called weft twill (Fig. 3). It is advisable to produce weft twill with a high weft density. If main overlaps prevail on the front side of the fabric, the twill is called main twill (Fig.4). It is advisable to weave the fabric in basic twill with a high warp density. For twill weave, a row purl is used.

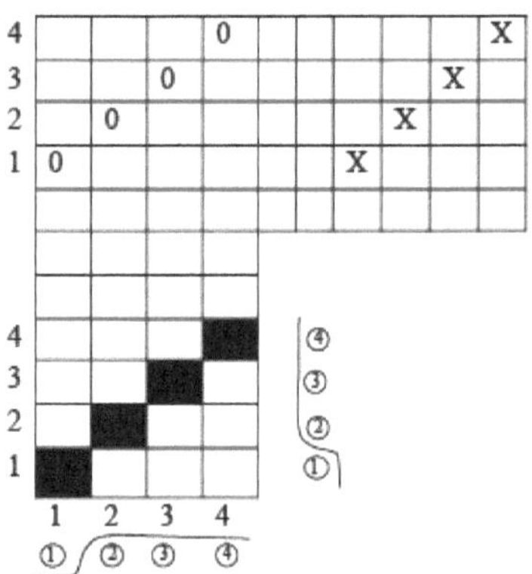

Figure 3. Full dressing pattern of twill 1/3.

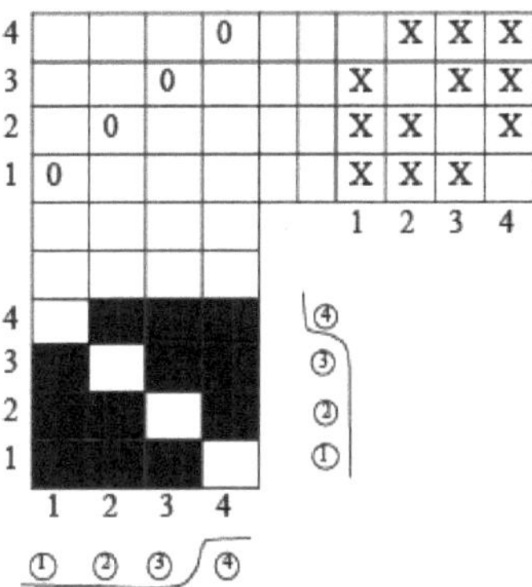

Figure 4. Complete filling pattern of a 3/1 twill.

The clarity of diagonals on the surface of fabrics depends on the following factors: warp and weft twist directions (a pronounced diagonal on the fabric surface is obtained by using different warp and weft twist directions); warp and weft thread thickness (the thinner the warp and weft threads, the more brilliant the diagonals on the fabric surface); warp yarn tension (increasing warp yarn tension in the production of warp twill will make the diagonals on the surface of the fabric less pronounced, while in the production of weft twill they are more pronounced and vice versa); the amount of overstitching (when working with overstitching, the diagonals on the surface of the weft twill fabric are more pronounced, while when working without overstitching, the diagonals on the surface of the warp twill fabric are more pronounced).

Satin (satin) weave

A characteristic feature of satin (satin) weave is that the weave of main and weft threads in the fabric is carried out by means of single main or weft overlaps, which are displaced relative to each other by the same amount of shift and evenly distributed within the weave pattern. Satin (satin) weave is denoted by a fraction, where the numerator shows the weave pattern **R**, and the denominator the shift **S**, and the numerator **R** and denominator **S** must not have a common divisor and for the shift the condition **1 S< < R - 1** must be observed. Satin weave forms long weft overlaps on the front surface of the fabric (weft density is higher than warp density), and single main overlaps are evenly distributed over the rapport area with a horizontal shift in the weft s_y. The satin weave forms long main overlaps on the front side of the fabric (the fabric density on the warp is higher than the density on the weft), and single weft overlaps are evenly distributed over the rapport area with a vertical shift on the warp s_o. To prevent yarn sliding in the fabric, the single, short overlap of each yarn should be placed closer to the centre of the long overlap of the previous yarn. For example, for a weave with $\mathbf{R} = 11$, it is appropriate to take the shift to be five or six. The construction of a satin weave is done in the following sequence:

1. On chequered paper draw a square with the number of cells equal to the rapports of the weave R_0 and R_y, number the main and weft yarns.

2. The first major overlap is marked at the intersection of the first sinkhole with the first base.

3. The second base slab is marked at the intersection of the second sink with the subsequent base, determined by the amount of horizontal displacement to the right in relation to the first slab.

4. The third base slab is marked at the intersection of the third sink with the subsequent base, determined by the amount of horizontal shear in relation to the second slab, and so on.

The satin weave is constructed similarly:

1. The first weft overlap is regarded as the starting point and is defined by the intersection of the first warp and the first weft.

2. The second weft overlap is marked at the intersection of the second base with the subsequent weft, determined by the amount of vertical shift upwards in relation to the first overlap.

3. The third weft overlap is marked at the intersection of the third base with the subsequent weft, determined by the amount of vertical shear in relation to the second overlap, etc.

Fig.5 shows the complete filling pattern of satin weave, and Fig.6 shows the pattern of satin weave constructed on the basis of satin 5/2 and satin 5/2.

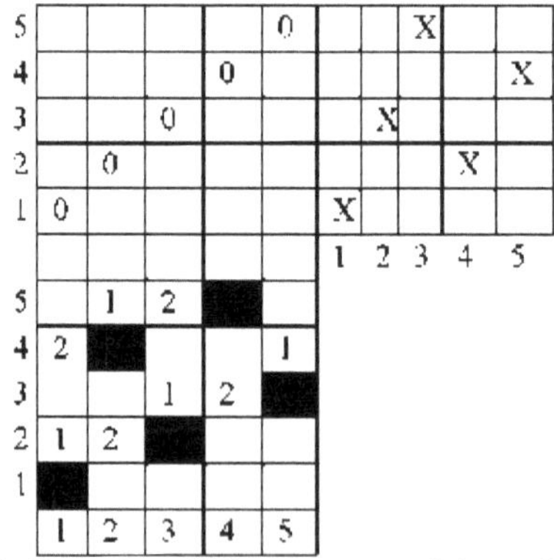

Figure 5. Complete filling pattern of the satin weave.

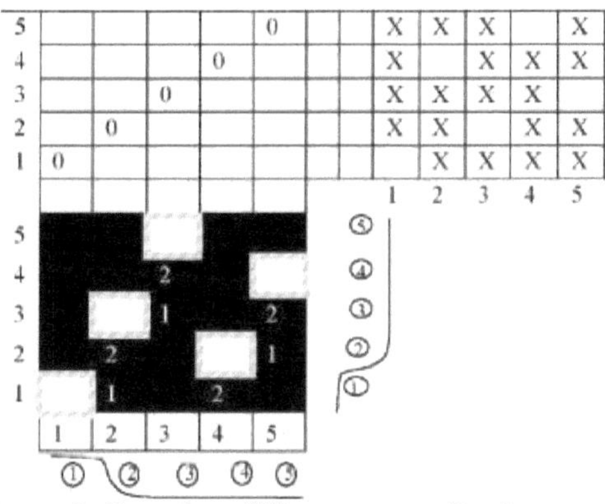

Figure 6. Complete filling pattern of satin weave.

As can be seen, the weave pattern $R_0 = R_{(y)} = 5$, and the horizontal shift $s_y = 2$ (for satin) and the vertical shift $s_0 = 2$ (for satin) is applied rowwise. The following points must be taken into account when designing and producing satin or satin weave fabrics:

-when the weave ratio increases, the fabric strength decreases and the fabric has a soft and shiny texture, and vice versa when the weave ratio decreases, the fabric strength increases and the surface of the fabric has a less smooth and shiny texture; to prevent striping on the surface of the fabric, the number of shifts should be close to half of the weave ratio, which ensures a uniform distribution of overlaps, as well as warp yarns should have a twist direction opposite to the direction of overlaps in the fabric;

-increasing the linear density of warp and weft forms a matt surface of the fabric, and decreasing the linear density of warp and weft forms a shiny surface of the fabric; increasing the density of the fabric on the warp (for satin weave) and on the weft (for satin weave) increases the lustre on the surface of the fabric and vice versa.

4.DERIVATIVES OF MAIN WEAVES

Derived weaves are fine-patterned weaves that retain the essential features of the main weaves and are formed by modifying the main weaves. As a rule, features of different weaves are not mixed in one pattern, which essentially distinguishes derivatives of main weaves from combined weaves. A distinction is made between plain weave, twill and satin (satin) derivatives. The plain weave derivatives are obtained by reinforcing single overlaps of plain weave yarns in the warp or weft direction or in the direction of both yarn systems. A basic rep weave is obtained by reinforcing a single overlap of plain weave yarns in the warp direction.

The **main rep** is denoted by a fraction, the numerator of which shows the value of the main planking (overlap), and the denominator the value of the weft planking (overlap), the main threads within the weave rapport.

The weft rapport is equal to the sum of the numerator and denominator, and the warp rapport to the rapport of the basic weave, i.e. plain weave. Fig.1 shows the complete filling pattern of the basic 2/2 rep, where $R_o = 2$, $R_y = 4$.

As can be seen from the cut of the fabric, the welts are located across the fabric, as the front and back surface of the fabric is covered with a dense deck of main threads, and the weft is not visible. If the density of the fabric on the warp allows to produce the fabric, the number of wefts in a dressing can be taken as two.

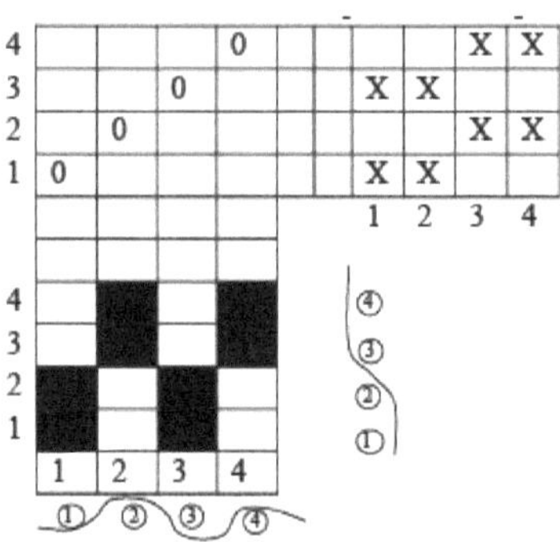

Figure 1. Complete filling pattern of the main rep 2/2.

The weft **rep** is denoted by a fraction, the numerator of which shows the value of the main planking (overlap) and the denominator the value of the weft planking (overlap) of the weft yarns within the weave pattern. The warp rapport is equal to the sum of the numerator and denominator, and the weft rapport to the basic (plain) weave rapport. Fig. 2 shows the complete filling pattern of a 2/2 weft rep, where $R_o = 4$, $R_y = 2$.

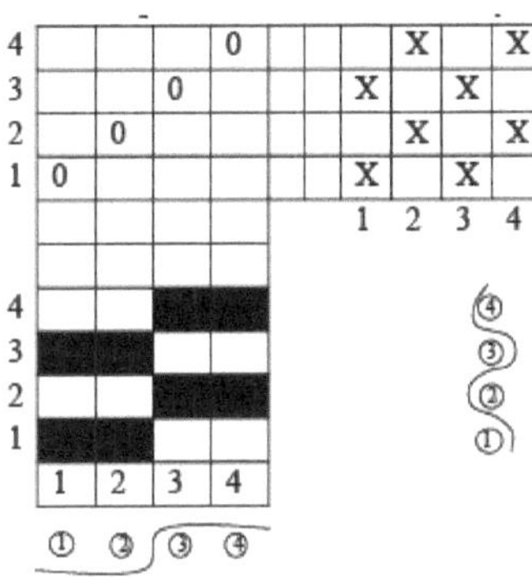

Figure 2. Complete filling pattern of weft rep 2/2.

The cut shows that the welts are located along the fabric, as the front and back surface of the fabric is covered with a dense weft yarn deck and the warp is not visible. For the production of this fabric, if the density of the warp allows it, the number of cuts in the dressing can be taken as two, using the corking in the cuts according to the pattern. If the overlap is increased, in the direction of only one warp of the weave rapport, we get a basic half-repeat. Fig.3 shows the complete filling pattern of the basic half-rep 3/1, where $R_o = 2$ and $R_y = 4$, i.e. the sum of the numerator and denominator.

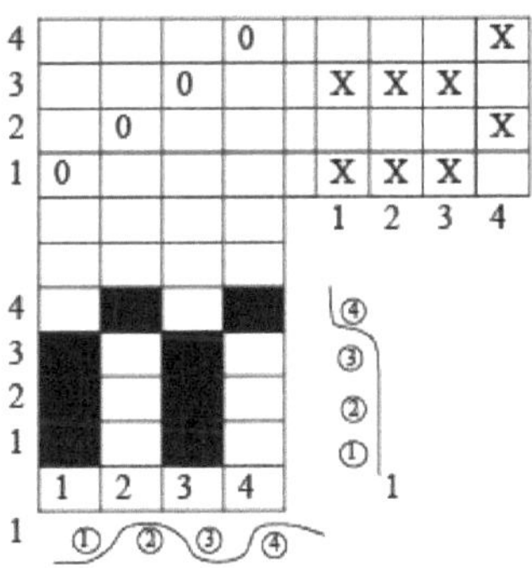

Figure 3. Complete filling pattern of the basic 3/1 half-repeat.

When reinforcing a single overlap in the direction of only one weft within the weave pattern, a weft half-repeat is obtained. Fig.4 shows the complete filling pattern of a weft half-repeat 3/1, where the sum of numerator and denominator is $R_o=4$, and the weft rapport is equal to that of the basic (plain) weave.

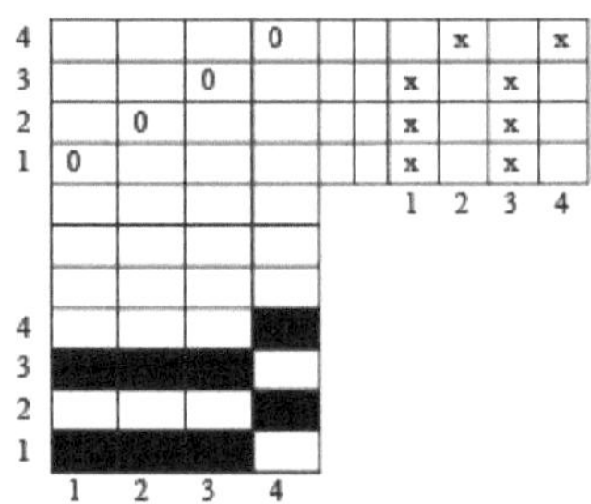

Fig. 4. Complete filling pattern of a 3/1 weft half-repeat.

The external effect of rep can be obtained in plain weave fabrics by the following methods: by a sharp difference in the linear densities of warp and weft yarns; by using two types of weft thick and thin. For this purpose it is necessary to have a multicolour device on the machine and weft laying should be carried out in the ratio 1:1 (peak-a-peak). This weft alternation creates protrusions (thick weft) and depressions (thin weft) on the fabric; differences in warp tension, e.g. even warp threads with high tension and odd warp threads with low tension, wound on separate warp threads.

When a single overlap in plain weave is lengthened in the weft and warp directions simultaneously, we get a plain weave. A plain weave is denoted by a fraction, with the numerator corresponding to the number of overlaps in the warp direction and the denominator in the weft direction. The sum of the numerator and the denominator is the rapport for the warp and weft. Fig.5 shows the complete filling pattern of a 2/2 bobbin, where $R_o = R_y = 4$. Cotte weave fabric has an effect in the form of small squares (checkers). The combination in one raportrait of cucumber and rape forms a shaped cotte. This makes it possible to obtain a greater variety of patterns with a different number of rapports, both on the warp and weft. Therefore, it is not possible to define the pattern of a shaped corduroy by any fraction.

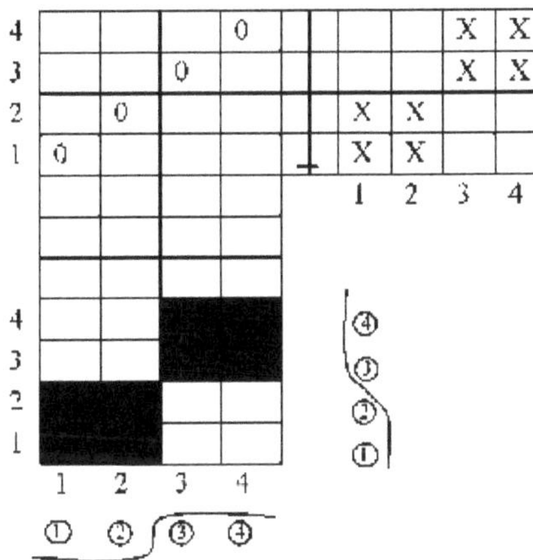

Figure 5. Complete filling pattern of the 2/2 cornet.

Fig.6 shows the filling pattern of a shaped cornet based on an irregular cornet with a basic rep.

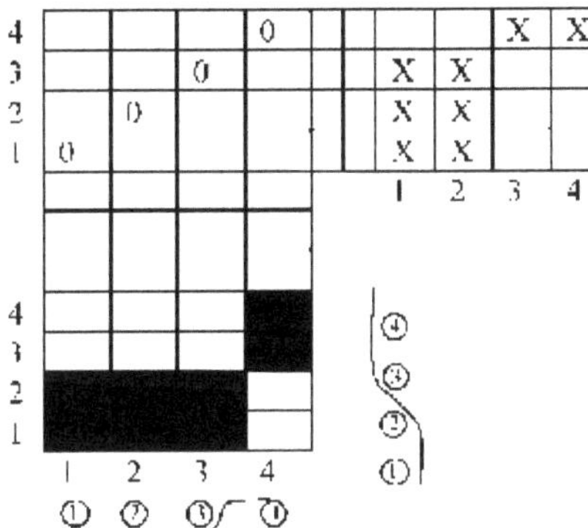

Figure 6. Filling pattern of shaped cornet based on irregular cornet with basic rep.

It is recommended to use the same linear density of warp and weft, as well as the same warp and weft density.

Twill weave derivatives

Twill derivatives are obtained by reinforcing single twill overlaps of the main twill weave, by changing the shear sign, by reinforcing overlaps and changing the shear sign. The derivatives of twill weave include reinforced twill, compound twill, broken , diamond (cross) twill, reverse-shifted twill, zigzag twill, shadow twill, etc. Reinforced twill is formed from plain twill by adding basic overlaps in the warp or weft direction.

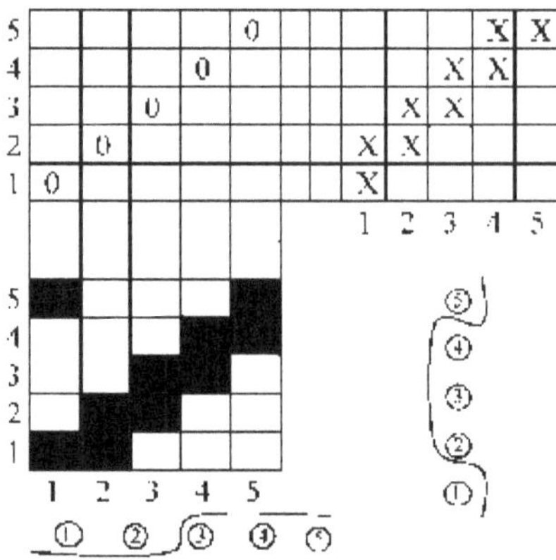

Figure 7. Filling pattern of reinforced twill 2/3.

Reinforced twill is denoted by a fraction, where the numerator is the number of main overlaps and the denominator is the number of weft overlaps. The sum of the numerator and denominator is the overlap ratio of the warp and weft. Reinforced twill can be with weft effect, where weft overlaps predominate on the front surface (Fig.7), or with basic effect, where basic overlaps predominate on the front surface (Fig.8).

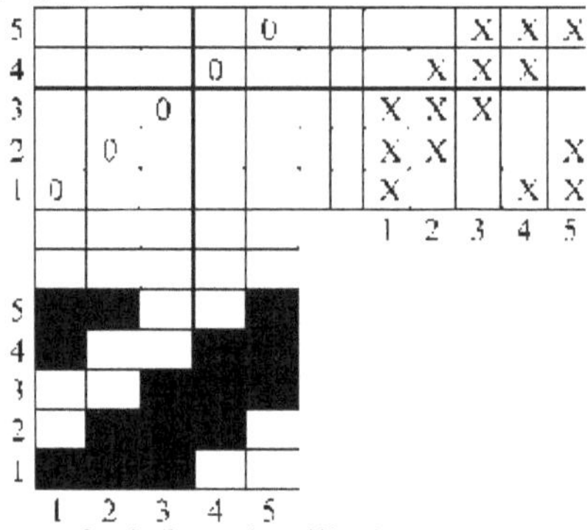

Figure 8. Filling pattern of reinforced twill 3/2.

If the number of overlaps (main and weft) is the same on both sides of the fabric surface (face and back), the twill is called double-sided (double-faced) (Fig.9), the difference between the front and back sides is only in the direction of the diagonal stripes.

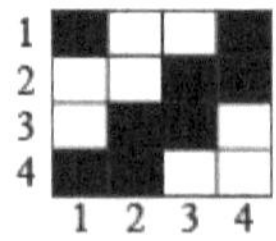

Figure 9. Weave of reinforced twill 2/2.

Fig.7 shows the complete filling pattern of reinforced twill 2/3, where $R_o = R_y = 2 + 3 = 5$, based on plain twill 1/4. Fig.8 shows the complete filling pattern of reinforced twill 3/2, where $R_o = R_y = 3 + 2 = 5$, built on the basis of plain twill 4/1. And on fig.9 fragment of weave of reinforced twill 2/2, where $R_o = R_y = 2 + 2 = 4$, built on the basis of simple twill 1/3. The reinforced twill forms wide and more distinct diagonal stripes on the surface of the fabric and has stronger warp and weft ligatures than in plain twill.

Complex (multi-band) twill is characterised by the presence of twill lines of different widths in a rapport or more. It results from the construction of two or more twill (simple or reinforced) weaves in succession. The rapport of a complex twill is indicated by several fractions, each of which constitutes a basic twill weave

$$R_o = R_y \, R_1 = + \, R_2 + \dots R_n$$

The sum of the numerator and denominator of a complex twill also makes up the warp and weft ratio of the fabric. Example, let's build a complex twill 1/2 2/2 (Fig.10), where the ratio on the base and on the weft is $R_o = R_y = 1+2+2+2+2 = 7$

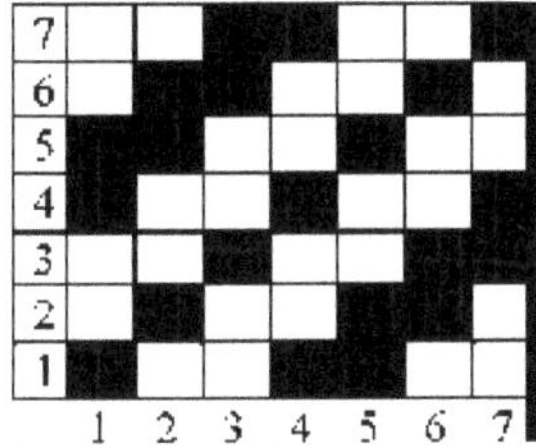

Figure 10. The weave of a complex twill 1/2 2/2.

The figure shows that for each weft yarn within the rapport, there is one main overlap, two weft overlaps, two main and two weft overlaps, the yarn shift is equal to one **S=1**. Complex twill can be weft, warp and double-sided depending on the number of overlaps on the front and back surfaces of the fabric.

A broken reverse twill is obtained by changing the shift sign of s_o or s_y from plus to minus after a given number of warp threads c_o or weft threads K_u. Changing the sign of the shift causes a change in the direction of the diagonal. The broken twill is constructed on the basis of any kind of twill weave, and the construction is made in the direction of the warp (broken on the warp) or weft (broken on the weft) and the tops of the teeth are located at the same level, i.e. the shift of the teeth is zero. Patterns are formed on the surface of the fabric in the form of longitudinal stripes, differing one from the other by the direction of twill diagonals, which differently absorb and reflect light rays, so some stripes seem dark and others light. Usually, the left part of the pattern (including the top of the prong) consists of a basic weave (has a straight diagonal that is directed from bottom left to top right), and the right part of the pattern (after the top of the prong) has a reverse diagonal that is perpendicular to the straight diagonal, with the reverse diagonals being shorter by two strands of the straight diagonals.

Weave pattern of broken twill on the warp

$R_o = 2\,C_o - \mathbf{2}, \quad R_y = R_6$

Broken twill on the duck

$R_y = 2\,K_u - \mathbf{2}, \quad R_o = R_6$

where: R_6 – rapport of the basic twill used for the construction of the broken twill; C_o, K_u – respectively the number of main and weft yarns, after which change the sign of the shift s_o and s_y from plus to minus.

A broken warp twill is obtained by changing the warp shift sign after the C_o of the warp threads of the basic twill, while keeping the weft shift sign. The order of building a broken twill on the base: set the base twill; set the number C_o after which change the sign of the shift on the base; calculate the rapport on the base and on the weft of the broken twill; build the weave pattern of the broken twill on the base. Example: to build a weave of broken twill on the base of twill 1/5 number $C_o = 6$, after which we change the sign of shift (Fig. 11) $R_o = \mathbf{2\,C_o\text{-}2} = 2 \times 6\,\text{-}2 = 10$, $R_y = R_6 = 6$. The purling in remise is reversed by six remise.

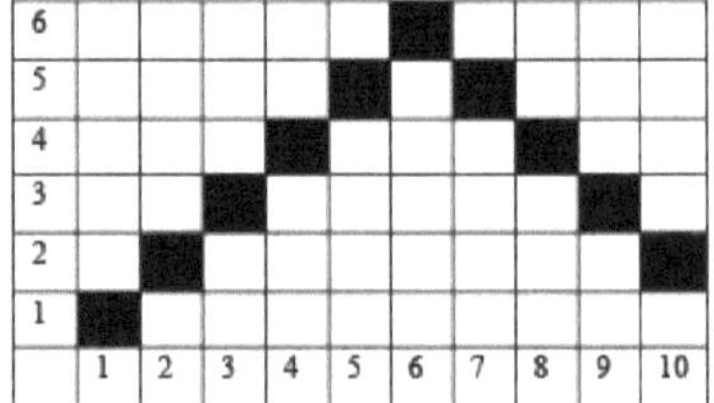

Figure 11. Interlacing of broken twill on a 1/5 twill base.

Example: construct the pattern of the weave of the broken twill on the base 1/4, with the number $C_o = 10$ (Fig.12). $R_o = 2\,C_o - \mathbf{2} = 2 \times 10\,\text{-}2 = 18 \qquad R_y = R_6 = 5$.

It follows from Fig.11-12 that the number C_o is the axis of symmetry, then the construction consists in the mirror image of the left part in the right part of the pattern. Punching in the remise is used in reverse or according to the pattern

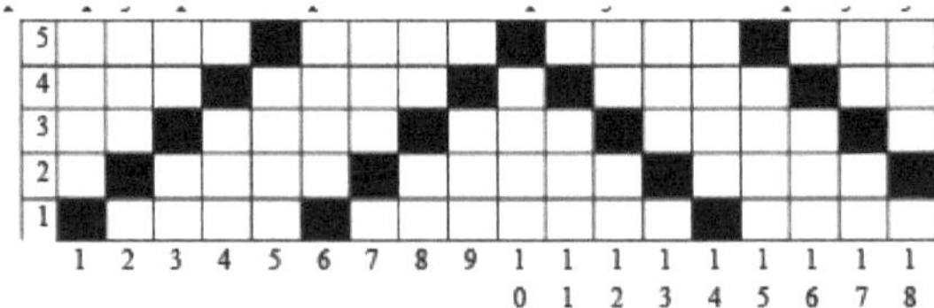

Fig.12. Interlacing of broken twill on a 1/4 twill base.

A broken weft twill is obtained by changing the weft shift sign after the K_u weft yarns of the base twill, while keeping the warp shift sign. Construction procedure: set the base twill; set the number K_u of weft yarns, after which change the sign of the weft shift; calculate the weave rapport; build a twill broken in the weft. Example: to build a twill broken in the weft on the basis of twill 1/7, the number $K_u = 8$, after which change the sign of the shift in the weft (Fig. 13). $R_y =$

$2\,Ku - 2 = 2 \times 8 - 2 = 14$. $Ro = R6 = 8$. We use a row warp purl for eight remise.

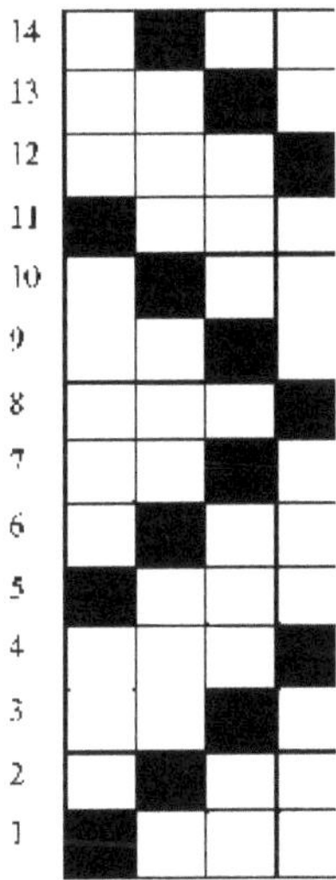

Fig.13. Weave of a weft broken twill on a 1/7 twill base.

Fig.14. Weft broken twill weave on a 1/3 twill base.

Example: build a broken weft twill on a 1/3 twill base, with $Ku = 8$, after which the sign of the weft shift is changed (Fig. 14). $Ry = 2\,Ku - 2 = 2 \times 8 - 2 = 14$. $Ro = R6 = 4$. The warp yarns are purfled in rows of four remise.

Rhomboid (cross-shaped) twill is obtained by changing the sign of the shift so from plus to minus after a given number of main threads co and changing the sign of the shift sy from plus to minus after a given number of weft threads Ku, i.e. the method of constructing a broken twill on the base and on the weft is applied simultaneously. The twill weave serves as a base for the construction and the rhomboid twill is denoted by the fraction of the base weave. Rhomboid twill can be basic, weft, equilateral, depending on the number of overlaps on both surfaces (face, back) of the fabric. The method of construction of rhomboidal (cross-shaped) twill: set the basic weave. determine the number of co and Ku, after which change the sign of the shift; determine R_o and R_y rhomboid twill R_o= 2Co - **2**, R_y= 2 Ku - **2** ; on the canvass paper build the basic weave, then from the upper end of the straight diagonal of the basic weave continue to build an additional three diagonals diverging in three directions at an angle of 90 ^{0} to each other within the rapport, and the length of the additional diagonals shorter by two threads of the straight diagonal. Fig.15 shows the complete filling pattern of rhomboidal weft twill 1/4 (Re=5), the number of threads after which change the sign of shift Co=5, Ku=5. Rapport of the rhomboid twill:

R_o= 2 Co - 2 = 2 x 5 - 2 = 8. R_y= 2 Co - 2 = 2 x 5 - 2 = 8. The purl of the warp yarns in the remise is reversed for five remises.

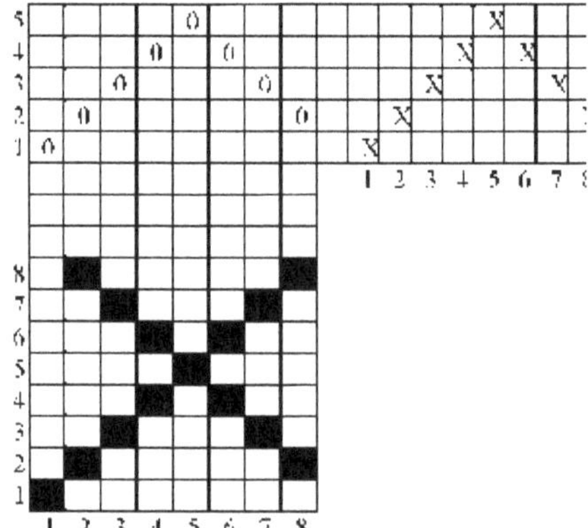

Fig. 15: Complete filling pattern of 1/4 diamond-shaped weft twill.

To increase the strength of the fabric, an additional overlap is placed in the centre of the rhombus, in our example we can place an additional main overlap at the intersection of the first main yarn and the fifth weft yarn.

Reverse shifted twill is a combination of right and left twill with the difference that when the shift sign is changed, the weave of the next threads is also changed, i.e. the main overlaps are replaced by weft overlaps and vice versa. Thanks to this, a clearer border between the weaves of the individual parts is formed on the surface of the fabric and the direction of the twill diagonals is different. Various twill weaves are used as base weaves and reverse shifted twills are constructed by increasing the number of threads on the base of the base weave by two or more times. There is a reverse shifted warp twill and a reverse shifted weft twill. The rapport of a reverse shifted twill is **R= 2K** in the break direction and **R**= R6 in the other direction.

Construction method: set the weave of the base twill; set the number **K** after which the diagonal breaks; calculate R_o and R_y; construct **K** threads of the base twill; complete the weave pattern. Example of construction, to build a tuck pattern reverse shifted twill on the basis of twill 1/5, with Co = 6. Rape on the base Ro=2Ko=2 x 6 =12, Ry= R6=6. In Fig.16 we build the base twill weave 1/5 on Co=6 threads, then change the sign of shift and transfer the overlaps from main to weft and weft to main in the right part of the figure. For this weave, we use a 12-repeat warp row purl.

Fig.17 shows the weave of a reverse shifted twill constructed on the basis of twill 1/3, number **K=8**, Ro= 2Ko=2 x 8 =16, R_y=4. The purl is used interrupted for 8 remise.

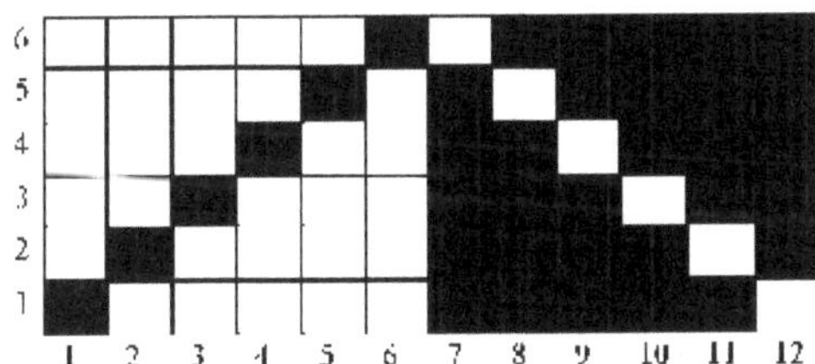

Figure 16. Weave of a reverse shifted twill warp on a 1/5 twill base.

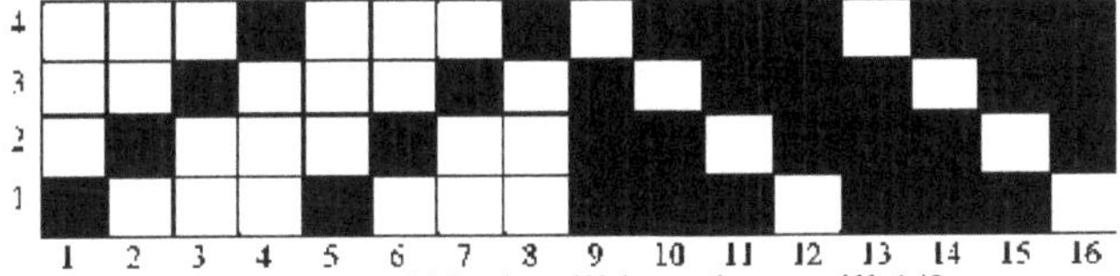

Figure 17. Weave of a reverse shifted twill based on twill 1/3.

Similarly, we build a reverse shifted weft twill (Fig.18-19). In Fig.18 the reverse shifted twill on weft is built on the basis of twill 1/5, with Ku = 6, Ro= R6= 6, Ry= 2Ku = 2 x 6 = 12, warp threads in remise row by row for 6 remise. For Fig.19 is built on the basis of twill 1/3, with the number Ku = 8, Ro= R6 = 4, Ry= 2Ku = 2 x 8 =16. The warp threads are picked in rows for 4 remise.

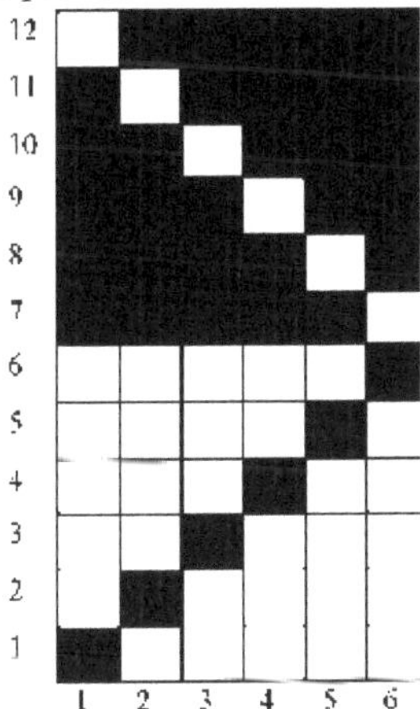

Fig. 18: Weave of a reverse shifted weft twill on a 1/5 twill base.

Figure 19: Weave of a reverse shifted weft twill on a 1/3 twill base.

Shadow twill is obtained by a gradual transition from weft twill to warp twill or vice versa. There are weft shadow twills and warp shadow twills. For a warp direction shadow twill, the warp rapport is $R_0=$ R6 x $\mathbf{K_{st}}$, where $\mathbf{K_{(st)}}$ is the number of steps in the shadow twill rapport $\mathbf{K_{st}}=$ R6 - **1**. Hence $R_0=$ R6 (R_e - **1**). The weft rapport $R_y=$ R_e. For shadow twill in the weft direction weft rapport $R_y=$ R_e (R_e - **1**). Rape on the base $R_0=$ R_e. Shadow twill fabrics are mainly produced on jacquard machines. Example: construct a shadow twill in the warp direction, on a 1/4 weft twill base, $R_e = 5$. Rapport on the warp $R_0=$ R_e (R_e - **1**) = 5 (5-1) = 20. Weft RAPPORT $R_y=$ $R_e = 5$. After determining the rapport $R_0 = 20$, and the number of steps $K_{st} = 5$, we draw on the canvass paper (Fig.22) the rapport of the basic twill $R_e = 5$, at that the basic weft twill 1/4 gradually passes

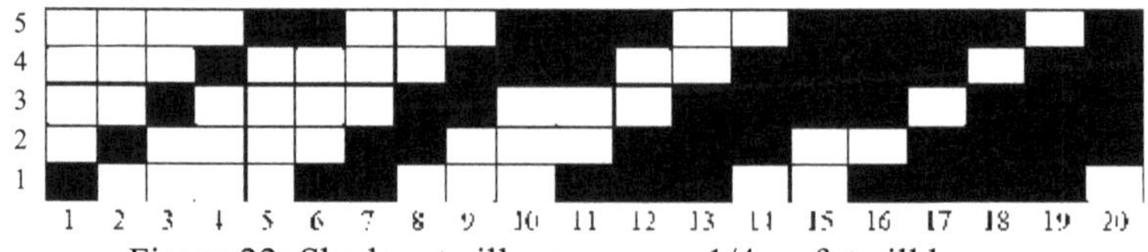

Figure 22: Shadow twill weave on a 1/4 weft twill base.

into the basic twill 4/1.

The warp threads are picked into the remise in a row, the number of remises corresponds to the number of warp threads in the rapport of the shadow twill weave.

Satin (satin) weave derivatives

Reinforced satins and satins are obtained by reinforcing single main overlaps (in satin weaves) and single weft overlaps (in satin weaves). **R>7, S>3** are used to construct reinforced satins and satins. Strengthening of single weaves determines the increased strength of fixation of threads in the fabric. The method of construction: we set and draw out the rapport of the basic satin. to single overlaps add one or several additional overlaps in any direction from the basic overlap; the number of additional overlaps should not exceed the value of the shift minus two; the rapport of reinforced satin (satin) is equal to the rapport of the basic weave. Example: to build a complete filling pattern of reinforced satin 7/3, on the basis of correct satin 7/3 (Fig.23). At first we build a regular satin with a rapport $R_o = R_y = 7$, with a shift **S = 3**. Then we introduce the first additional basic overlap at the intersection of the first utochina with the second basic, the second overlap is defined by the second utochina with the fifth basic, the position of which is determined by a shift equal to three, etc.

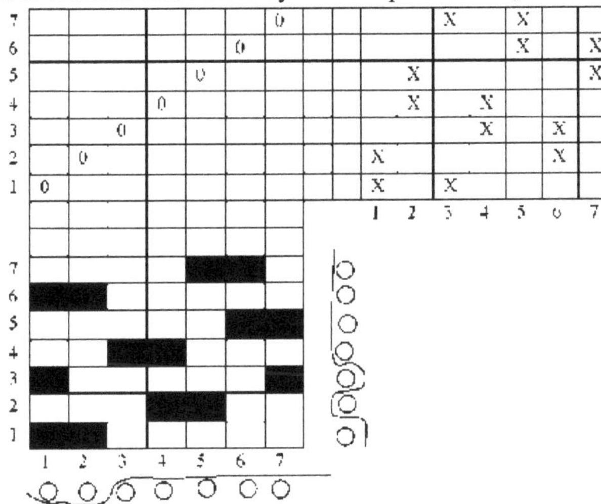

Fig. 23: Full filling pattern of reinforced satin 7/3 based on satin 7/3.

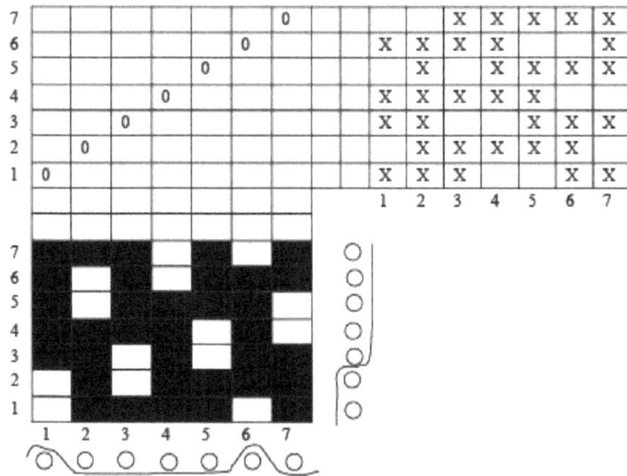

Figure 24. Full filling pattern of the reinforced atlas 7/3 based on the atlas 7/3.
Fig.24 shows the complete filling pattern of the reinforced satin 7/3 based on satin 7/3. In both cases, the purl is rowed and the number of hems is equal to the base weave.

Reinforced satins are used for high weft densities, reinforced satins for high warp densities.

Shadow satins (satins)

Shadow satins (satins) are obtained by a gradual transition from weft overlaps to main overlaps or from main overlaps to weft overlaps. These weaves are used mainly in jacquard fabrics similar to shadow twill. There are shadow satins (satins) in the weft direction and shadow satins (satins) in the warp direction. For the shadow satin (satin) in the direction of the warp, the base rapport $R_o=$ R6 (R_e - **1**), where, **R6** is the rapport of the base satin (satin). The weft rapport $R_y=$ R_e. For shadow satin (satin) in the weft direction, the weft rapport $R_y=$ R_e (R_e - **1**). Rappage on the base $R_o=$ R_e. Example: to build a shadow satin in the direction of the base on the basis of satin 5/2 (Fig.25), the rapport of the base satin $R_o=$ $R_y=$ 5, the overlap shift is two **S** = 2. The rapport of shadow satin $R_o=$ R_e (R_e - **1**) = 5 (5-1) = 20. The weft rapport $R_y=$ R_e = 5. In Fig.25, we construct the first step (warp threads 1-5) of basic (correct) satin 5/2, the second step (warp threads 6-10) of reinforced satin, the third step (warp threads 11-15) of reinforced satin, and the fourth step (warp threads 16-20) of basic (correct) satin 5/2.

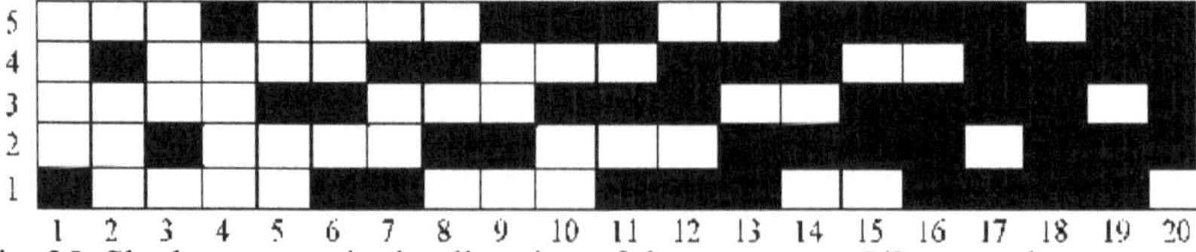

Fig. 25: Shadow sateen in the direction of the warp on a 5/2 sateen base

Irregular satins (satins)

Irregular satins (satins) can be obtained with minimum rapport **R=4**, and the rapport and shift have a common divisor, such satins (satins) include fabrics with rapport **R=4**, R=6, etc. For the construction of such satins (atlases) the overlap shift is variable. Example: to build a four-thread satin $R_o = R_y = 4$, the shift is taken variable **S=1,2,3**. Fig.26 shows the complete filling pattern of irregular satin 4/1,2,3, and Fig.27 of irregular satin 6/2,3,4.

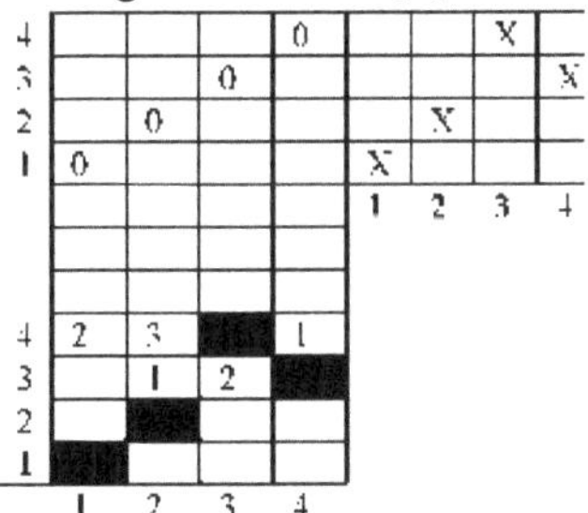

Fig. 26: Complete filling pattern of irregular satin 4/1,2,3.

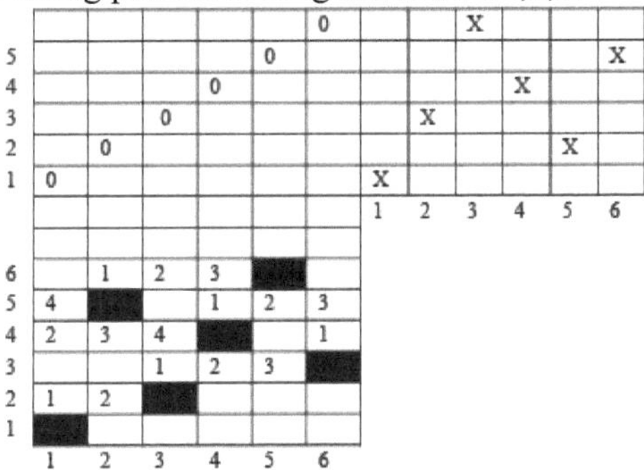

Figure 27. Complete filling pattern of irregular satin 6/2,3,4.

5.COMBINED WEAVES

These weaves are characterised by patterns of different shapes and character on the surface of the fabric, which are formed by the combination of main and derived weaves. Combined weaves are constructed in such a way that they can be produced with dobbies. The formation of combined weaves is possible by the following methods: combination of separate types of main and derived weaves; permutations of single or group threads (warp, weft); selection of coloured threads in combination with the weave pattern (for coloured patterns). Combined weaves are differentiated into weaves with striped or checked patterns, waffle, diagonal, crepe, translucent, with fixed flooring (welt), with coloured patterns.

Weaves with striped and chequered patterns

These weaves are based on the main and derived weaves and form longitudinal and transverse stripes, as well as cells, which are located across the entire width of the fabric. To enhance the effect on the fabric surface, the following are used: left and right, regular and shaped twist yarns; two-colour yarns (curled yarns with coloured threads); special reeds with different tine densities according to the weave pattern; special with variable densities of fabric diversion mechanisms. Strip weaves can be along the warp (longitudinal strip) and across the warp (transverse stripes). The weave pattern depends on the width of the strips, the density of the fabric and the type of strip weave. For a longitudinal stripe weave, the warp rapport is equal to the sum of the main yarns in the base weaves that make up the strips $R_o = P_{01} + P_{02} + ... P_{op}$, where: $n_{oi}, P_{o2}, ... P_{op}$ is the number of basic threads in each strip. The number of main threads in each strip must be a multiple of the rapport of the base weave of the strip

$n_{oi} = P_{oĭ} \cdot a_1$; $P_{o2} = P_{oo2} \cdot a_2$; $P_{op} = P_{oη} \cdot a_η$. Where: $P_{oi}, P_{o2},$. **Poi** - density of the fabric on the base in strips, thread/cm; a_1, a_2, a_p - width of the strips forming the weave rapport, cm. The weft rapport is defined as the smallest common multiple of the weft rapports of the weave strips.

For the production of fabrics with longitudinal stripes (Fig. 1), a consolidated interrupted warp picking in remise is used, where the number of vaults is equal to the number of stripes in the fabric of different weaves. The production of fabrics with longitudinal stripes is possible with the same values of workmanship in each strip, which is achieved by the same number of overlaps with equal yarn rapports of each strip. Otherwise, it is necessary to use a multi-sewing filling for the warp yarns of each strip.

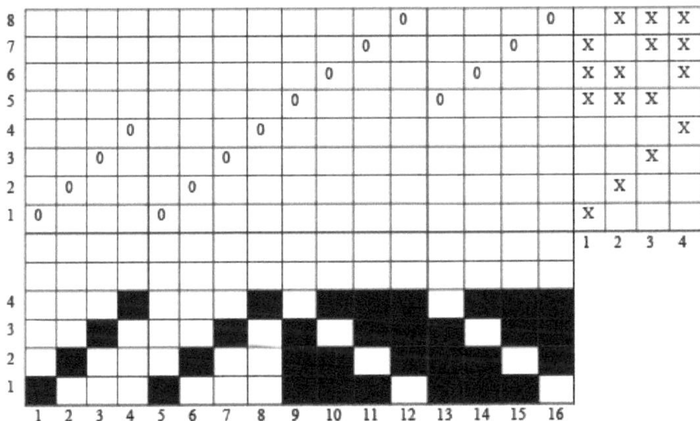

Figure 1. Full dressing pattern of the longitudinal stripe fabric.

In order to clearly separate the stripes on the border of the transition, different overlaps are placed (opposite warp-weft or opposite weft-base) or additional (coloured, different thickness, type, etc.) threads are placed. Example: to build a fabric with two longitudinal stripes $R_{o1}= R_{o2=20}$ threads/cm, width of stripes $a_1= a_{2=0},4$ cm, weave of the first strip weft twill 1/3, and the second strip twill 3/1. This variant of weave gives the possibility to produce the fabric on the machine with single warp threading. Fig.1 shows the complete filling pattern of the longitudinal strip fabric based on twill 1/3 and 3/1. Rapport of weave of fabric in longitudinal stripe on the base $R_o=$ $P_{0\check{i}\text{-}\alpha_1}+$ **P_{o2}-$\alpha2$** $= 20 - 0,4+20 -0,4=16$. The weft rapport is the smallest common multiple of $R_{yi}=$ 4, $R_{y2} = 4$, R_y 4. Thread picking in the remiz discontinuous summary, the number of threads picked in the reed tooth is four.

For a cross-strip weave (Fig. 2), the weft pattern is the sum of the weft yarns in the strips forming the weave pattern

$R_y=$ $n_{yi}+ n_{y2}+...$ P_{oop}, where: $P_{u1}, P_{u2},... P_{up}$ - number of weft yarns in each strip. The number of weft yarns in each strip should be a multiple of the weft rapport of the basic weave of the strip $n_{yi}= P_y{}^{\wedge} e_i$; $P_{u2}= R_u{}^{\wedge}v2$; $P_{oop}= R_u{}^{\wedge}vp$. Where: R_u - density of fabric on weft, thread/cm; c_1, **c2**, c-width **of the** strip forming the rapport of the weave, cm. The base rapport is defined as the smallest common multiple of the rapport on the base of the strip weave.

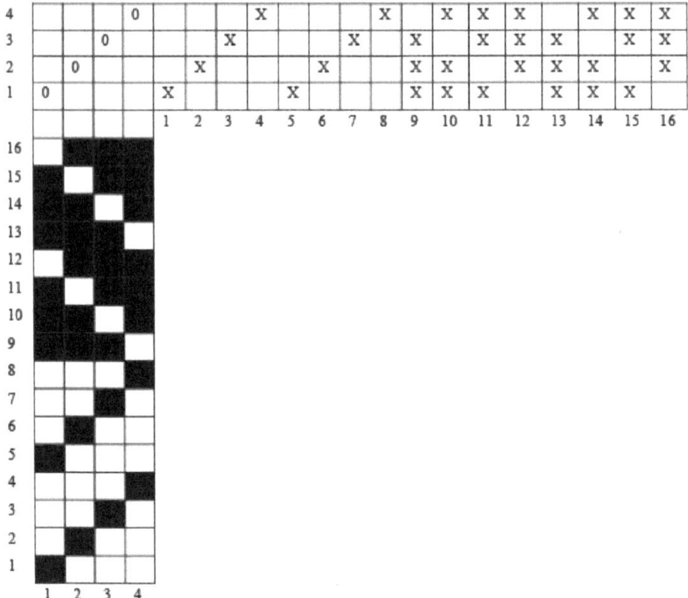

Figure 2. Full dressing pattern of cross-striped fabric.

When weaving fabrics with transverse stripes, row striping is used. Example: to build a fabric with two cross stripes $Ru_1= Ru_2 = 20$ N/cm, the width of the stripes **in 1= in2** = 0,4 cm, the weave of the first strip weft twill 1/3, and the second strip main twill 3/1. Rapport of the weave of the fabric in cross stripes in weft $Ry= Ru^{v1}+ \mathbf{Ru^{v2}}= 20\text{-}0,4+20 - 0,4=16$. The rapport on the base is equal to the smallest common multiple $Ro_{i=4}, Ro_{2=4}, Ro_{=4}$. The number of remizoks in the threading is equal to the base pattern, i.e. four. The threads are picked through the heddles in a row.

For plaid weave, the weave pattern depends on the size of the cells, the density of the warp and weft, and the type of weave in the cells (Fig. 3).

Determination of the weave ratio of checked fabrics is made by the same methods as for determining the weave ratio of striped fabrics.Rape ratio on the warp $Ro= Po_{i\text{-}\alpha_1}+ Po_{2\text{-}\alpha_2}+...+ Po_{\eta\text{-}\alpha\eta}$. Weft rapport $Ry= Ro^{v1}+ Ro^{v2}+..... .+,.I\backslash\text{-}Bии$. Example: construct a weave in a cell, $Ro_1=Ro_2=Ru_1=Ru_2=20$ threads/cm, width of the cell **a1= a2= c1= c2** = 0,4 cm, weave of the

the first cell twill 1/3, the second cell twill 3/1. Rapport on the base of the fabric in the square R_o= $_{Poi - ai}$+ $_{Ro2}$ - $\mathbf{a2}$ = 20 -0,4+20 -0,4=16. Rapport on the weft of the fabric in the cell R_y= $_{Ru^v1}$+ $\mathbf{Ru^v2}$= $\mathbf{20-0,4+20-0,4=16}$.

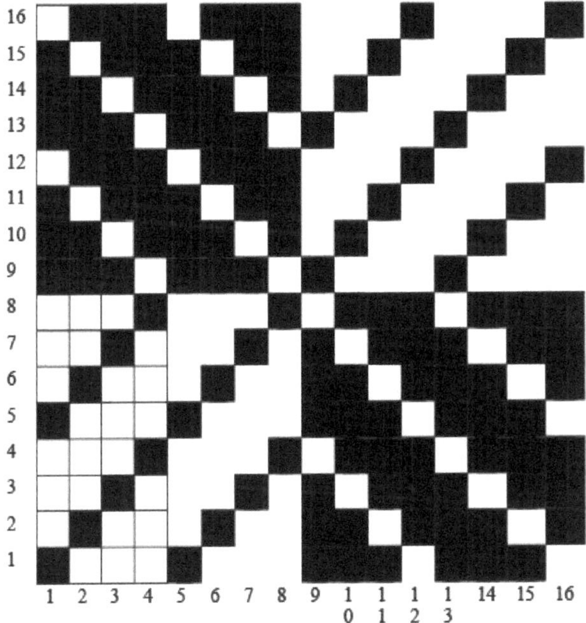

Figure 3. The weave of the fabric in the cage.

weave of a checked fabric. An analysis of the filling pattern of a checked weave will show that on the surface of the fabric we have a pattern in the form of four squares (two with weft effect, two with main effect), the size of each square being equal to the size of the basic weave. On the border of the squares against the basic overlaps are weft overlaps, and against the weft overlaps are basic overlaps. The yarns in the remise are interrupted and consolidated.

Waffle weaves

The waffle weave forms a relief pattern on the surface of the fabric in the form of cells with convex sides and a concave centre. The convex sides are formed by long main and weft overlaps, and the concave centre by short main and weft overlaps (plain weave). The relief (expressiveness) of the pattern depends on the linear density of the yarn, the density of the warp and weft, and the basic weave. The waffle weave is constructed on the basis of a lozenge twill derived from plain twill. Rapport of the weave on the base R_o= $_{2Ko}$ - $\mathbf{2}$, on the weft R_y= $_{2Ku}$ - $\mathbf{2}$, where: $_{Co}$ $_{Cu}$ - respectively the number of warp or weft yarns, after which there is a change in the sign of shift (Fig. 4).

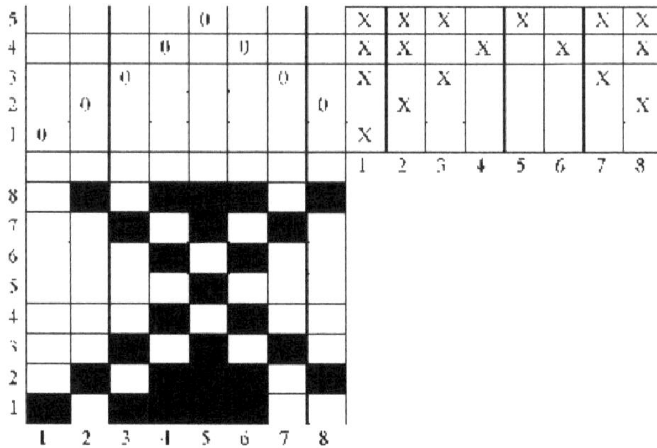
Figure 4. Complete filling pattern of a waffle weave.

Methodology of waffle weave construction: a rhomboidal twill is constructed; the inner part of the rhombus is filled in staggered order with basic planks, leaving one row of weft overlaps up to the contour of the rhombus. Example: to build a waffle weave on the base of rhomboid twill 1/4 number of threads, after which, the sign of shift changes, $C_o = K_{u=5}$. Rapport on the base $R_o = 2K_o - 2 = 2\text{-}5\text{-}2 = 8$, on the weft $R_y = 2K_u - 2 = 2\text{-}5\text{-}2 = 8$. Further, having put off on the kanvovy paper (fig.4) the rapport on the base and on the weft, we build a rhomboidal twill, and then having departed from the rhombus on one weft overlap we complete the long basic planking (overlaps). Proborku take reverse, on five remise, ie the number of remise $K_r = 5$. Also waffle weave is built on the basis of a complex twill, the rules of construction are the same, in this case the fabric has a beautiful appearance. A small twist of yarn contributes to better moisture absorption when producing towelling fabrics.

Diagonal weaves

Diagonal weaves form a relief pattern on the surface of the fabric in the form of slanting stripes, similar to a twill weave, directed from bottom left to top right. The angle of inclination of the diagonal depends on the warp density and the density ratio of the fabric.

If $R_o = R_u$, then the diagonal angle $\alpha = 45^0$, if $R_o > R_u$, then $\alpha > 45^0$. The warp density should not exceed twice the value of the weft density ($2P_u$), otherwise there is irrational use of raw materials and unequal strength of the fabric. A further increase of the diagonal angle is possible by increasing the warp shear (vertical shear) more than one $S_o > 1$. Therefore, a twill weave having a warp shift of more than one ($S_o > 1$) is called a diagonal weave and a twill is called a diagonal weave. For diagonal weaves, the slope angle (α) is determined $\mathbf{tga} = P_o S_o / P_y$, where: Po, R_o- the density of the fabric in warp and weft; S_o- the shear in warp (vertical). The base is a complex twill, which has pronounced diagonals from the main overlaps; wide - for greater relief and narrow - to prevent sliding of the fabric.

Two options are possible when constructing a diagonal weave:

1 .When dividing the basic weave rapport of a complex twill weave by the increased warp shift. The warp ratio $R_o = R6/S_o$. Weft ratio $R_y = R6$.

2 .When not dividing the basic twill weave pattern by the increased warp shift. Rape on the warp and weft $R_o = R_y = R6$. Example: to build a diagonal weave on the base of a complex twill 4/1, 3/2, vertical shift $S_{o=2}$, diagonal weave pattern on the base $R_o = R6/S_o = 10/2 = 5$. Weft pattern $R_y = R6 = 10$.

In Fig. 5a, let $R_{o=5}$, $R_{y=10}$. Then, for the first warp thread, we mark four main overlaps, one weft, three main and two weft overlaps, i.e. $4+1+3+2 = R_y$. For the second warp thread, having stepped back two steps (shift $S_{o=2}$), we will also place four main, one weft, three main and two weft slabs. Similarly, with a shift, we build the overlaps of the third main thread, etc. Marking the points of the beginning of the main overlaps for each main thread, we can clearly see a steep diagonal. To make a diagonal, five remizas are required, i.e. the number of remizas $K_p = \mathbf{R6/So}$ for row sashing. Example: to build a diagonal weave on the base of a complex twill 4/1, 4/2, shift $S_{o=2}$, the rapport of the diagonal weave is indivisible by shift, so $R_o = R_y = R6 = 4+1+4+2 = 11$. In Fig.5b we mark $R_o = R_{y=11}$. Then for the first warp thread, we set the overlaps to 4(main) + 1(weft) + 4 (main) + 2(weft). For the second main thread, having stepped back two steps ($S_{o=2}$), we build similar main and weft overlaps in the same order, etc. By marking the beginning points of the main overlap of each warp yarn, we see a steep diagonal.

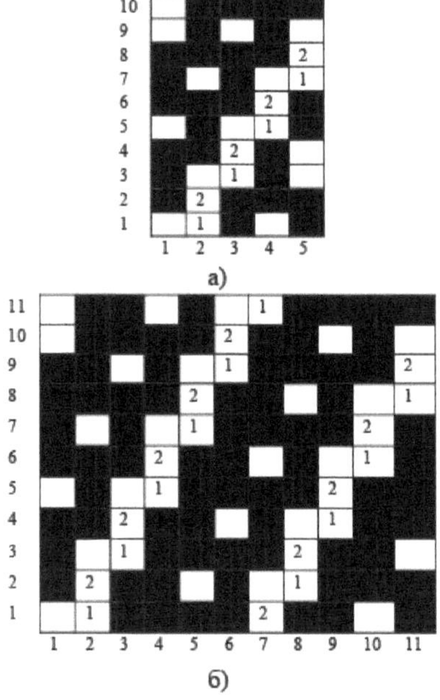

a)

б)

Figure 5. Diagonal weave.

For this weave, the number of heddles in the threading is equal to the basic weave ($K_P = R_{(6)}$), i.e. eleven heddles for a warp row.

Crepe weaves

Crepe fabrics are fabrics that have a crepe (fine-grained) effect on the surface due to the twist of the threads. Such fabrics include crepe de chine, crepe georgette, crepsatin, etc. Krepdeshin produced from the main threads of silk - raw and weft threads crepe right and left twist (2200 kr / m), and crepe georgette (3200 kr / m). It is also possible to imitate the crepe effect on the fabric with appropriate weaves. For correct construction of crepe weaves it is necessary to observe the following requirements: on the surface of the fabric there should be no visible stripes or clearly defined patterns; there should be no large groupings of weft or main overlaps; on the surface of the rapport there should be approximately equal number of weft and main overlaps. Crepe weaves are constructed on the basis of main weaves or their derivatives. There are the following methods of crepe weave construction:

adding basic or weft overlaps; overlapping (combining) two or more weaves; placing threads of one weave between threads of another weave; rearrangement of threads of the same weave; rotation of the basic weave; negative, i.e. replacing basic overlaps with weft overlaps and vice versa.

In the method of adding basic or weft overlaps, the patterns of the basic weave are broken. For example: the basic weave is accepted four rapports of plain weave (Fig.ba), $R_o = R_y = 8$. The rapport of the crepe weave is equal to the rapport of the basic weave. Violation of the pattern is carried out arbitrarily (Fig. 6b) due to the addition of basic overlaps in two directions at the intersection of the first warp of the second weft, the second warp and the third weft, the third warp and the fourth weft, the fifth warp and the eighth weft, the sixth warp and the seventh weft, the seventh warp and the sixth weft, in compliance with the requirements of crepe weave construction.

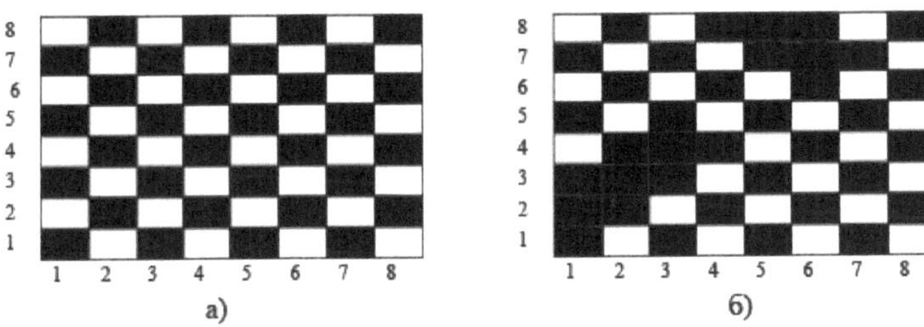

Figure 6. Crepe weave.

The overlapping method consists of taking the same number of threads or their multiples in the rapports of two basic weaves and overlapping the rapports of the basic weaves, with overlapping overlaps removed. Example: Draw up a weave pattern based on plain and six-strand irregular sateen (Fig. 7).

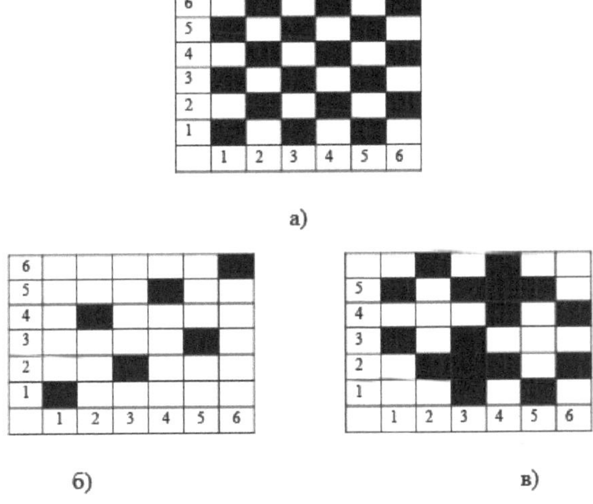

Figure 7. Crepe weave.

We construct plain weave with repetition of the rapport three times, i.e. R_o= R_y=6 (Fig.7a), six-strand irregular satin with R_o= R_y=6 (Fig.7b). Then we superimpose plain weave (Fig.7a) on six-strand satin (7b) and in the places where the main overlaps overlap, we remove these overlaps (Fig.7c), i.e. the first warp is the first weft, the second warp is the fourth weft, the fifth warp is the third weft, the sixth warp is the sixth weft. The weft rapport R_y= $R6K$.

When the weft yarns of one weave are placed between the weft yarns of another weave. The weft rapport is R_y Ki/Ku, where: Ku- the alternation of the weft yarns of one weave in relation to the weft yarns of another weave, which is the sum of these weft alternations. The base rapport R_o= $R6K$. Example: construct a crepe weave by placing the warp threads of plain weave between the warp threads of twill 1/2 , the alternation of warp threads is 1:1. Rapport of plain weave weave is equal to two R_o= R_y=2 (Fig.8a), and twill 1/2, R_o= R_y=3 (Fig.8b). THE total smallest multiple of the basic weave rapport $R6K=6$. Consequently, the weft rapport of crepe weave R_y= $R6K=6$. And the rapport on the base of crepe weave

$$R_o= R6^\wedge K_y = 6\,(1+1) = 12$$

In Fig. 8c, let's set aside the rapport of crepe weave equal to Ro=12 and R^ó. Then put in Fig.8c on odd warp threads 1,3,5,7,9,11 overlaps plain weave, and on even threads 2,4,6,8,10,12 overlaps twill weave. For this weave, we adopt a four-remise pattern striping or twelve-remise row striping.

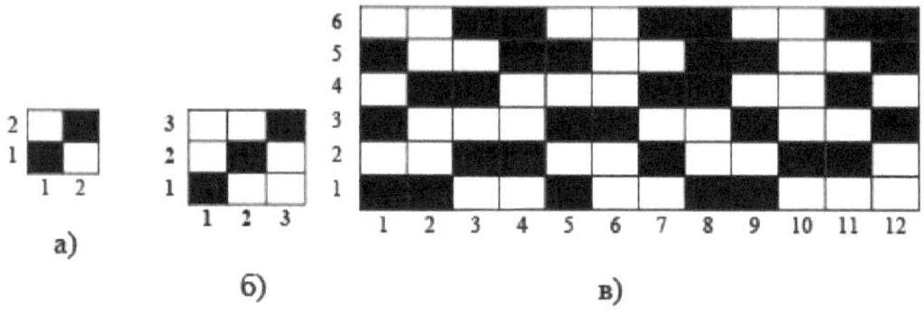

Figure 8. Crepe weave.

Example: construct a crepe weave by placing the weft yarns of plain weave between the weft yarns of weft rep 2:2, weft yarn alternation 2:1. The rapport on the basis of the plain weave basic weave is two (Fig. 9a), and the rapport on the basis of the basic weave $R6K=4$. Therefore, the pattern on the base of the crepe weave R_o R= $^\wedge$ =4. The weft rapport of plain weave and rep weave is two, therefore, the base weave rapport $R6K=2$.

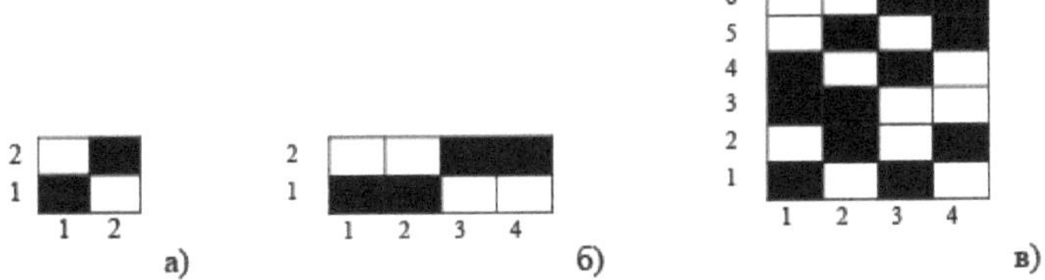

Figure 9. Crepe weave.

Since the alternation of weft yarns is 2:1, the sum of these ratios is three, i.e. Ku = 3. Consequently, the weft ratio of the crepe weave Ry Ibk Ku= 2-3 = 6. In Fig.9c, let Ro=4 and Ry=6. Then on weft wefts 1 and 2 we will put plain weave overlaps, and on the third weft weft we will put basic overlaps (for the first weft in Fig.9b), as the ratio of wefts is 2:1. Then on the 4th and 5th we put plain weave overlaps on the 4th and 5th weave and on the 6th weave basic overlaps on the 6th weave (for the second weave Fig.9b). We accept a row-by-row purl on four remise.

A method of rearranging the threads of the same weave to break the pattern of the basic weave. The rearrangement is done with main or weft yarns, single yarns or groups of yarns. The pattern of the crepe weave is equal to the pattern of the basic weave. The yarns are picked into the remise in rows. Example: Create a crepe weave on the base of a complex twill 1/2, 2/2 by rearranging the warp threads.

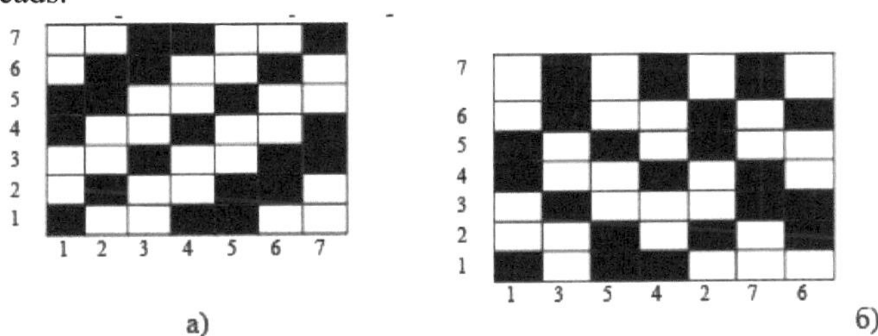

Figure 10. Crepe weave.

The rapport on the warp and weft of the crepe weave is equal to the basic weave, i.e. Ro= Ry= R6=7. In Fig.10a we plot the basic weave of twill weave 1/2, 2/2. Next to it (Fig.10b) we mark the rapport of crepe weave and at the same time we rearrange the warp threads of the basic weave in the following sequence 1,3,5,4,2,7 and 6. The crepe weave is constructed similarly by rearranging the weft threads (Fig.11b) on the basis of the base weave of the complex twill 1/2, and 2/2 (Fig.11a). Here we rearrange the weft threads in the following sequence 1,3,7,2,6,5,4. The rearrangement is done arbitrarily, in compliance with the rules (requirements) of crepe weave construction.

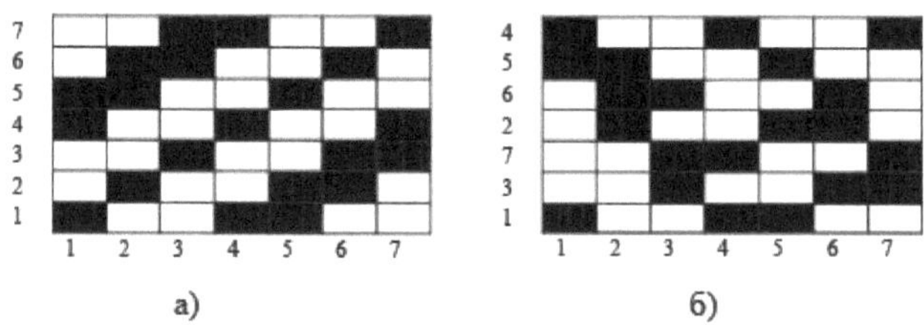

a)   6)

Figure 11. Crepe weave.

The rotation method consists of transferring the base weave from one quarter square to the other in a clockwise direction, while rotating the pattern 90^0 anti-clockwise. The weave rapport is equal to twice the value of the base weave. In the warp Ro= 2R6. In weft Ry= 2R6. The yarns are picked into the remise in a row, for the number of remises equal to **K**= 2R6. Example: to build a crepe weave using the rotation method on the basis of an arbitrary pattern (Fig. 12a) with the rapport Ro= Ry=3. Using the rotation method, rotate 90^0 counterclockwise Fig. 12a and get Figure 12b, then rotate Figure 12b in the same direction at $90^{0,}$ get Figure 12c, similarly get Figure 12g.

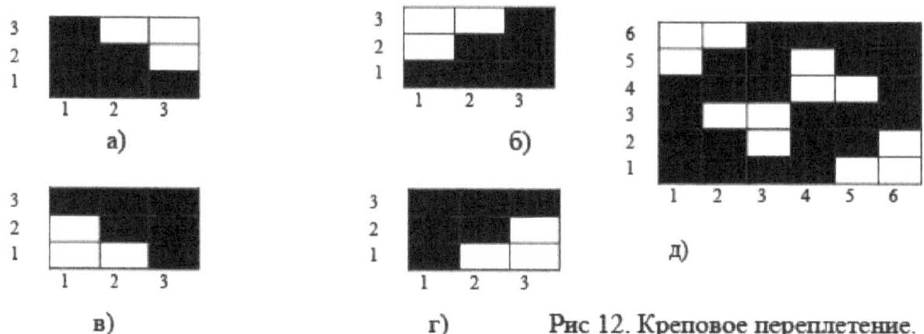

а)   б)
в)   г)   Рис 12. Креповое переплетение.

Figure 12: Crepe weave.

Then in the first quarter of the square with a base and weft ratio of six (Ro= Ry= **2R6**= 2-3 = 6) Fig.12d. Let's place (weave) Fig.12a and sequentially in each quarter of the square clockwise weaves Fig.12b, 12c, 12d.

The negative method consists of transferring the basic weave from one quarter square to another clockwise, with simultaneous rotation of the pattern 90^0 clockwise and changing the overlap from basic to weft or from weft to basic. The size of the crepe weave is equal to twice the size of the basic weave: Ro=2R6, Ry= 2R6. Example: to build a crepe weave by negative method on the basis of an arbitrary pattern (Fig.13a) with the rapport Ro= Ry=3. The initial position of Fig.13a, we make a turn at 90^0 clockwise, with a simultaneous change of overlap from the basic to weft, and weft to basic (Fig.13d). Turning 90^0 clockwise of the weave Fig.13b with simultaneous change of overlaps from weft to main and main to weft we get the weave Fig.13c. Similarly, we get the weave of Fig. 13d. In Fig.13d, in the first quarter of the square we will place the weave of Fig.13a, in the second quarter the weave of Fig.13b, in the third quarter of Fig.13 and 6 quarters of the square the weave of Fig.13g. Analysing the crepe weave shows that the second quarter has a reflection of the first quarter as a negative, the third quarter as a negative, the fourth quarter as a negative and the first quarter as a negative.

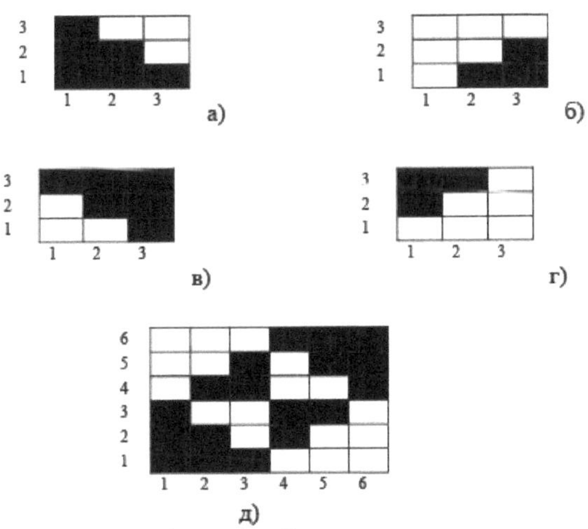

Figure 13. Crepe weave.

There are no restrictions in the construction of crepe weaves, so it is possible to construct crepe weaves by any method, i.e. using several methods (placement, superimposition, rotation, etc.) with observance of the rules of construction of crepe weaves. Such construction determines the obtaining of a beautiful appearance of the fabric, due to the increase of the weave rapport.

Translucent weaves

They are obtained by combining long and short overlaps. Gaps are formed when warp and weft yarns are arranged in rows with sharply different planking lengths. Particularly noticeable gaps in the combination of plain weave and rep weave, the longer the planking (overlap), the higher the effect of translucency. Method of weave construction: set the rapport of the translucent weave R_o and R_y; determine the number of threads in the warp and weft (p_o, p_y) $n_o = n_y = R/2$; construct the translucent weave. Example. to build a translucent weave with a base and weft rapport $R_o = R_y = 10$, for this we use a plain weave and a basic rep 5/5 (Fig.14) and the number of threads in the bundle $P_o = P_u = 5$.

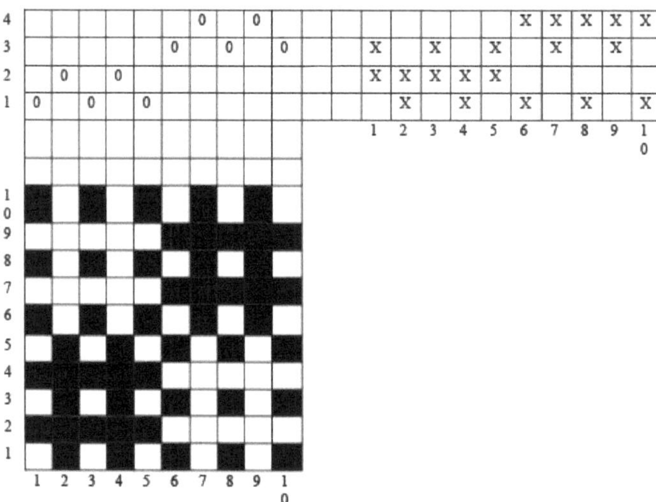

Figure 14: Translucent weave.

The warp threads are picked into the reed in a bundle of five threads, and the warp threads are picked into the remise according to the pattern for four remises.

Rubbing weaves

It is produced on the basis of weaves with long main or weft plaids (overlaps). On the front side of the fabric longitudinal or transverse - convex stripes (rubs) are formed. As basic weaves (with long main or weft overlaps) it uses basic or weft 4/4, 5/5, 6/6, 8/8, twills 4/4, 6/6, 8/8 and others. The fabrics based on these weaves have a loose structure, so fixing weaves with short overlaps, such as plain or twill 1/2 or 2/1, are introduced. When selecting a fixing weave, it is necessary that the rapport be equal to or multiple of the number of threads overlapping the planking. The rapport of the rubbing weave when securing long main overlaps (longitudinal rubs) $R_o= R_{o6}$ $\mathbf{R_{(H}}$; $R_y= R_{y6}$. Rapport of welt weave when securing long weft overlaps (cross welts) $R_o= R_{o6}$; $R_y= R_{y6} R_{y3}$, where: R_{o6}, R_{y6} are the rapport of the base weave in the warp and weft respectively; R_{03}, R_{y3} are the rapport of the setting weave in the warp and weft respectively. Example: Create a filling pattern for a cross welt weave fabric based on a 6/6 weft rep. The fixing weave is taken plain (twill 1/2 is also possible). We determine the rapport of the transverse welt weave on the weft and on the base $R_y= R_{y6}- R_{y3}=2 - 2=4$, $R_o= R_{o6}=12$. In Fig.15 we put the main overlaps in the direction of the weft of weft rep 6/6 in the rapport with a twist twice. Then, on the continuation of the first weft on the second welt, mark the short basic overlaps of the plain weave (warp yarns 7,9,11). For the second weft, on the first ruching, with a shift S=1, form short basic plain weave overlaps (warp yarns 2,4,6).

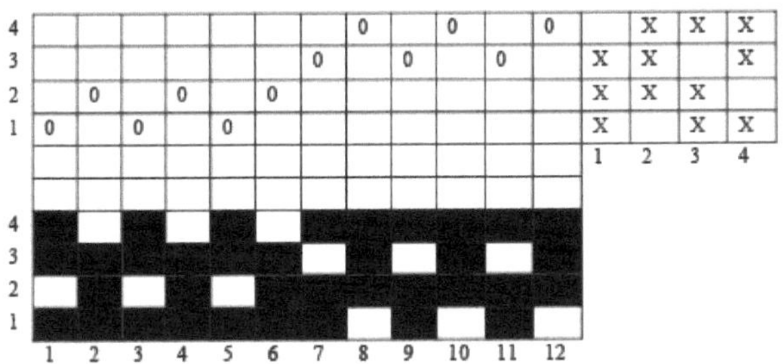

Fig. 15: Cross welt weave.

For the third point, form short plain weave overlaps (warp yarns 8,10,12) on the second welt with a shift S=1. On the fourth weft, similarly to the second weft, we form short plain weave overlaps, observing the shift **S=1**. The thread picking in the remise is free - interrupted for four remises. Example: to construct a filling pattern of a longitudinal welt weave fabric on the basis of the basic rep 6/6. The fixing weave is accepted plain weave. We determine the rapport of longitudinal welt weave on the base $R_o= R_{o6} -K_{o3}=2-2=4$, on the weft $R_y= R_{y6}=12$. In Fig.16 we build the basic rep 6/6 with repetition within the rapport $R_{o=4}$. Then on the second ruching of the first warp yarn we determine the short overlaps of plain weave (these are weft yarns 7,9,11). For the second warp yarn on the first scar we determine the short overlaps of plain weave (weft yarns 1,3,5), observing the shift of plain weave **S=1**. The short overlaps for the third and fourth warp yarns are constructed in the same way, observing the conditions of plain weave shift.

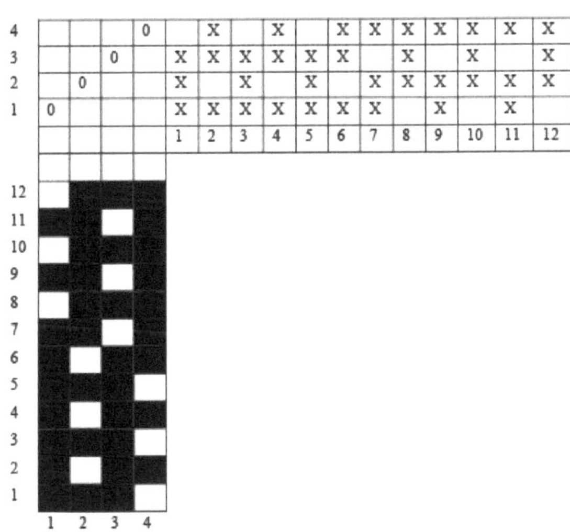

Figure 16. Longitudinal welt weave.

We use a row of four remise.

Weaves with coloured pattern

A pattern is obtained on the surface of the fabric by using coloured warp and weft yarns. The ratio of a weave with a coloured pattern is equal to the smallest common multiple of the basic weave ratio R_6 and the colour ratio Itz. Colour rapport is the number of basic or weft coloured threads, after which the order of alternation of colours is repeated. Methodology of weave construction: determine the basic weave rapport: determine the colour rapport; determine the rapport of the colour pattern; depict the weave of threads in the fabric, with dots mark the main overlaps; paint the main and weft overlaps with paint of the colour that have warp and weft threads.

6.PLAIN JACQUARD FABRICS

Simple jacquard fabrics occupy a large place in the range of jacquard fabrics. These fabrics are constructed with two systems of yarns: one system of main yarns and one system of weft yarns. In the design of single-layer fabrics good results are achieved by applying shadow developments of pattern details, successfully combining raw materials in the warp and weft, as well as by using different combinations of raw materials in one of the thread systems. Wefts of different raw materials lie side by side in the fabric, in the same plane, thus preserving the single layer of the fabric. Often one high number weft forms the background of the fabric, and another, usually low number, continuing the weave of the background, forms the pattern. The fit of the wefts may be even or odd. The range of single-layer jacquard fabrics includes dress and blouse fabrics, headscarves, lining fabrics, blanket satins, tablecloths, decorative fabrics, gastuchny fabrics. Depending on the nature of the pattern of the fabric to be tucked in, on the number of rapports along its width, as well as depending on the density of the fabric at the usual jacquard tucking, the pool board is divided into parts equal to the number of parts of tucking, or rapports of piercing. Each part of the pool board has as many holes as the number of working hooks in the jacquard machine and as many arcade cords, and therefore, the main threads in one rapport of purl. The peculiarity of jacquard machines, that, hooks have a directed movement from each other and each hook carries out lifting and lowering (movement) of one warp thread (in the presence of the width of the filling all the different intertwining warp threads) and several warp threads (in the presence of the width of the filling repeating patterns of different intertwining parts of the warp threads). The hooks are connected with arcade cords, threaded into the cassette board and connected to the faces, in the eyes of which the warp threads are threaded. Hangings or elastic cords are tied to the faces. The width of the cassette board is equal to the width of threading on the reed, and the number of holes in the board is equal to the number of harness cords. When threading the machine it is necessary to know the position of the first hook (the first warp thread), which is determined by the position of the prism relative to the Jacquard head according to the rule of the left hand. In Fig. 1a, the prism is on the right side of the Jacquard head. Therefore, if you stand facing the prism, the left hand will indicate the position of the first hook and the right hand will indicate the position of the last hook (2400 hook in our example). Consequently, according to the left hand rule, the first hook will be the farthest left hook and the last hook will be the closest right hook (2400 hook). To prevent crossing of the arcate cords, the arcate cord threaded into the first hole of the cassette board is tied to the last hook (2400 hook), and the arcate cord threaded into the last hole of the cassette board is tied to the first hook. In this case, the pattern on the fabric has the opposite direction to the pattern on the chuck. will be the closest right arcate cord threaded into the first hole of the cassette board is tied to the first hook. In this case the direction of the pattern on the fabric coincides with the direction of the pattern on the chuck. Such a direction of the pattern on the fabric has the prism arrangement above the breastplate (Fig.1c). In this case, the first hook is in the far left corner and the last hook in the near right corner. When the prism is placed on the left side of the jacquard head (Fig.1b), the first hook will be the farthest left hook and the last hook (2400 hook). Fig.1g shows the prism arrangement over the base, where the first hook is at the near right corner and the last hook at the far left corner. The pattern on the fabric has a different direction from the direction of the pattern on the chuck because the arcuate cord threaded into the first hole of the cassette board is tied to the last hook (2400 hook), and the arcuate cord threaded into the last hole of the cassette board (2400 hole) is tied to the first hook. To obtain the same pattern direction on the fabric and on the chuck in the threading variant (Fig.1a and 1g), the fabric is worked face down on the machine. In the jacquard head, the hooks are arranged in longitudinal and transverse rows, which depend on the size (number of hooks) and division (large, medium and small) of the jacquard machine. Usually, the number of hooks in the cross row in large division machines is 8 rows, medium division - 12 rows, small division - 16 rows. Dividing the total number of hooks by the number of cross rows gives the number of longitudinal rows in the jacquard head

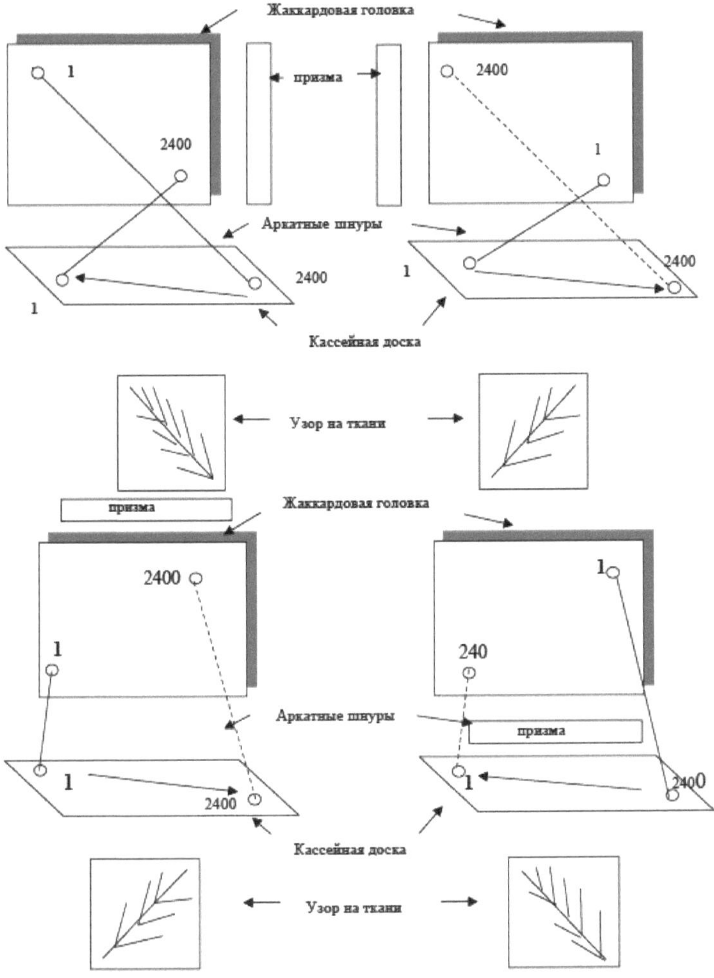

Fig 1. Prism arrangement on jacquard machines.

The pool board distributes the harness cords evenly across the width and depth of the loom insertion and is positioned between the Jacquard head and the warp threads. The number of holes in the cassette board corresponds to the number of harness cords and the number of warp cords. The number of holes in the cross row equals the number of hooks in the cross row of the jacquard head. The holes are arranged in staggered order. The first hole (Fig. 1a and 1g) is always the furthest left hole and the last hole is always the closest right hole (also for multi-part threading). Depending on the nature of the pattern, the number of pattern rapports on the width of the fabric the pool board is divided into parts or on the rapport of the strip and to each working hook tie as many arcate cords as there are parts in the dressing. Width, one part of the dressing is determined by dividing the number of working hooks on the density of the fabric on the basis. There are the following types of corks arcate cords in the casseinny board - row, reverse, combined, summary. Row corking is used in the production of fabrics with asymmetrical pattern. If the pattern is repeated once along the width, the striping is called single-part row striping (Fig.1a and 1g), where the number of hooks $Kk=2400$, the base rapport $Ro<2400$, the number of threads in the base $Po<2400$. In multi-private row picking, several arcuate cords are controlled by one hook. Fig.2 shows the scheme of two-part row picking, where the number of hooks $Kk=2400$, the pattern on the base $Ro=2400$, and the number of threads in the base $Po=4800$

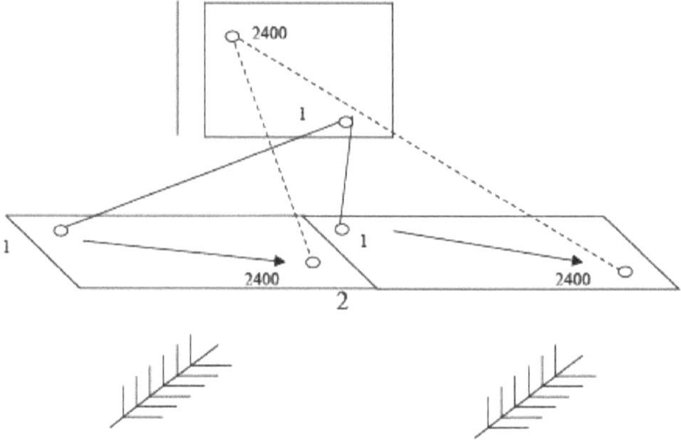

Fig.2. Two-part row piercing.

The number of pieces in a filling is allowed not more than 16. Backstitching is used in the production of fabrics with a symmetrical pattern on the warp. Backstitching is divided into single-part and multi-part. At single-part reverse picking (Fig.3) the number of working hooks $Kk=2400$, the base pattern $Ro=4800$, the number of threads in the base $Po=4800$.

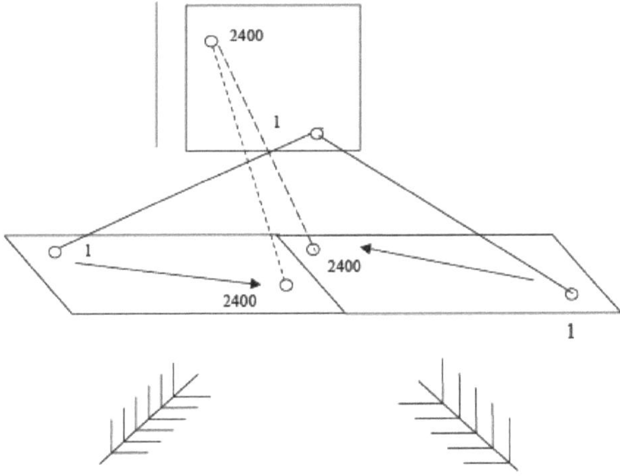

Figure 3. One-part reverse punching.

At two-part reverse stripping (Fig. 4) the number of working hooks $K_k=2400$, the rapport on the base $R_o=4800$, the number of threads in the base $P_o=9600$ threads.

Combined fringing is used for piecework with a background and border. Since the border is symmetrical, a reverse fringe is selected for it. For the background, a row fringe is selected. In Fig.5, the border is reverse purfling and uses 1 to 400 hooks, while the background uses row purfling from 401 to 2400 hooks.

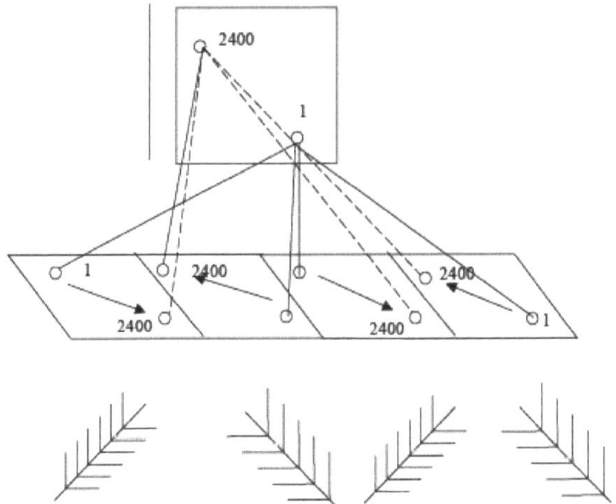

Figure 4. Two-part reverse punching.

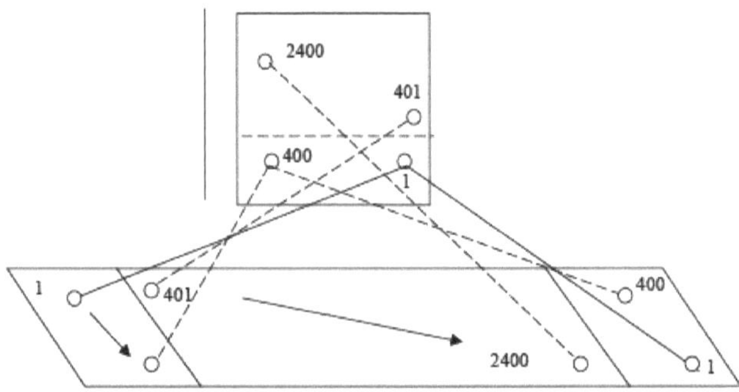

Figure 5. Combined furring.

Joint punching in the production of complex large-patterned fabrics with more than two thread systems (tapestry, carpet, etc.). In this case, the hooks on the Jacquard head and the cassette board are divided into vaults by depth and each vault controls one main thread system. Fig.6 shows a two-vault single-part row selvedge.

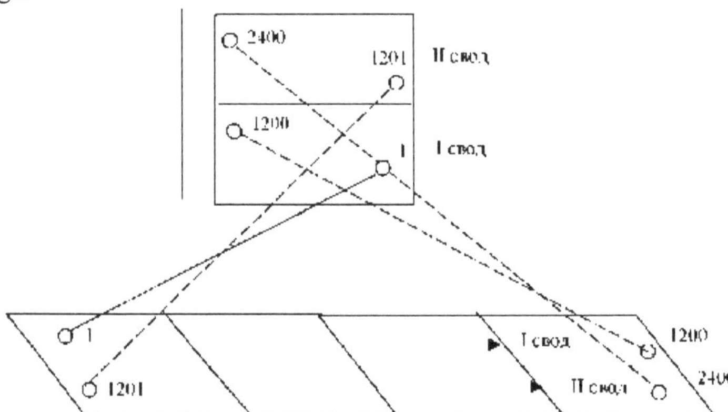

Fig 6. Two-wire single-part single-part row purlin

Patronising in jacquard weaving

The representation of the weave of a jacquard fabric on canvass paper is called patronising. Patronising can be of three types: firstly, new patterns for fabrics produced in industry; secondly, patterns for new fabrics based on existing loom dressings; thirdly, patterns for new fabrics based on new loom dressings. In the first type of patronising, the density of the warp and weft fabrics, the type of raw material, the number of hooks in operation, the type of arch cords, the width of one piece and the total width of the threading, etc. must be known and the same types of weave are used as in the production of the previous fabric patterns. In the second type of patronising, the number of working hooks, filling density in warp and weft, the size of the piercing rapport, the type of piercing and the total width of the filling are known. Known filling parameters are used in the technical calculation. The third type of patronising involves a new filling. In this case, the number of working hooks, the new filling density in the warp and weft, the type of fraying of the arcade cords into the cassette board are selected, then the technical calculation of the fabric is given. The method of patronising depends on the structure of the fabric, cardboard die-cutting machines, on the qualification of the insetters. There are the following methods of patronising: - patronising in unfolded form with full application of weave; -patronising in unfolded form with partial application of weave; patronising in reduced form. In unfolded patronising with full weave application, each vertical row of the canvass paper corresponds to one main yarn and the horizontal row to one weft yarn. This method is the most labour-intensive, as the weave is applied over the entire area of the patron and is used in single-ply fabrics with several weaving effects, as well as in some multilayer fabrics (one and a half-ply and two-ply). Usually, a painted small cell on the canvass paper corresponds to the lifting of the main threads (main overlap) and hence to the lifting of the hook, and the lifting of the latter is provided by a hole cut in the cardboard. When making the cartridge, all main overlaps are painted in one colour and the holes on the cardboard are cut accordingly to this cardboard colour. When patronising in unfolded form with partial weaves, the weave is applied to the pattern area, and one or more rapports (one or two large cells) are depicted on the background. This method is used for patronising single-layer fabrics with two weaving effects - background and pattern. Patronising in a reduced form is used when producing multilayer fabrics consisting of several systems of warp and weft yarns. This method is the fastest and easiest, as the weaves are not applied to the cartridge, but painted in as many colours as there are effects (patterns) on the fabric. Model weaves are developed for each chuck colour and applied to the chuck. The disadvantage of this method of patronising is the difficulty of notching the cards, because when notching the cardboard the cartridge is read as many times as the number of model weaves used. The development of the cartridge is carried out in three stages. At the first stage specify the filling parameters: size, type and division of the jacquard machine; the number of working and auxiliary hooks; distribution of working and auxiliary hooks on the machine; type of picking arcate cords into the cassette board; position of the first hook; number of parts in the filling; width of one part of the filling and the total filling width of the fabric; the size of the pattern rapport on the base, the characteristics of the fabric. At the second stage the following operations are carried out: choose the method of patronising; calculate the canvass paper; calculate the patron; specify the number of weaving effects - choose the weaves for each effect and the method of fabric production (face up or face down). At the third stage: execute tracing paper; execute the chuck; draw up instructions for card notching. R_{on} is the number of vertical rows in the cartridge or the number of small cells in the cartridge on one horizontal.

The warp pattern is determined by multiplying the warp density of the fabric R_o by the horizontal pattern size **a**. Ku **Ro-a** (1).

R_{yn} is the number of horizontal rows in a pattern or the number of small cells in a pattern per vertical. The weft rapport is determined by multiplying the weft density of the fabric, R_u, by the vertical pattern length **in.**

$R_{yn}=$ $I\backslash^B$ 2). The weft pattern must be divided into the background and selvedge weave pattern and the weft change pattern.

In unfolded patronising, the base and weft pattern pattern rapports are equal to the base and weft pattern rapports. $R_{on}= R_{oy3}$ (3). $R_{yn}= R_{yy3}$ (4).

The number of coarse cells in the pattern (K_{ok}, K_{uk}) is determined by the ratio of the pattern's base pattern ratio to the number of vertical rows of the pattern and, respectively, the pattern's weft pattern ratio to the number of horizontal rows of the pattern of one coarse cell, i.e.

$$K_{ox} = R_{on}/K_o \qquad (5)$$
$$K_{yx} = R_{yn}/K_y \qquad (6)$$

Patronisation of jacquard fabric is carried out in the following sequence: allocate the pattern layout from the drawing; transfer the pattern outline to the tracing paper; determine the border and size of the pattern layout on the tracing paper; calculate the number of threads in the pattern layout on the warp and weft; determine the number of hooks of the Jacquard machine; calculate the canvass paper (by the number of small and large cells); divide the tracing paper into cells; transfer the pattern outline to the canvass paper; apply the pattern weave on the canvass paper; check the patron. Selection of the pattern outline from the pattern given for patronising is carried out according to the fabric sample developed by the artist. Then the outline of the pattern is transferred to tracing paper. The borders of the pattern outline are marked with straight lines in the warp and weft directions. The values of the pattern spread on the base and on the weft are determined by the formulas (1) and (2). The number of small cells horizontally (c_o) is equal to the number of hooks in the cross row of the machine, the number of small cells vertically is determined by formula (2). The number of large cells horizontally K_{ok} and vertically K_{uk} by formulae (5) and (6). Division of tracing paper into large cells is made in horizontal and vertical direction (Fig. 1b). The designation of cells is carried out horizontally (from left to right), vertically (from bottom to top) the contour of the pattern on the tracing paper is transferred in the enlarged scale. Binding is applied on the elements of the pattern accordingly to the structural and shadow effects.

The chuck check determines the correct arrangement and transition of weaves at the joints of pattern elements, taking into account structural contrast or shadow effect in individual pattern sections. The filling calculation of the jacquard fabric is supplemented by the calculation of the arcade cords and the cassette board.

Number of arcate cords in the bundle $A_{,,}$ $P_f/K^{\wedge}$, where: P_f is the number of background threads in the filling; R_{oy3} is the rapport of the warp pattern across the width of the fabric. The width of the cassette board is determined by $v_k = v_{zb} + (1 \% 2)$ cm, where: v_b is the width of the filling on the reed, cm.

Total number of holes in the cassette board. $O_o = P_f + P_{kr}$, where: P_{kr} is the number of threads in the selvedge.

Number of parts in the cassette board $K_{.,}$ P_f/K_{uz}

The number of holes in the cross row of the cassette board depends on the density of the main yarns in the fabric. At higher densities the number of holes is equal to the number of hooks in the cross row or a multiple of the number of hooks in the cross row, and at lower densities the number of holes is half the number of hooks.

Number of cross rows (holes) in one part of the cassette board

$N = R_{oy3}/nK.$

Number of cross rows (holes) per 1 cm $N_{1k} = \mathbf{Ro/Pk.}$

Hole density on the cassette board $R_{ock} = \mathbf{Oo/Vk^{\wedge}Pk.}$

When filling fabrics of medium yarn counts, the number of holes per 1 cm should not exceed 5 holes. If the density is higher, increase the number of holes per cross row (P_k).

LITERATURE

1. Rakhimkhodjaev S.S., Kadyrova D.N. Theory of tissue structure. Textbook. Tashkent. Adabiyot uchkunlari. 2018. - 212 pp.

2. Damyanov G.B. et al. "Fabric structure and modern methods of fabric design" - Moscow, Legkaya Industriya 1984g

3. G. H. Oelsner's. A Handbook of Weaves is the best known and most accessible weaving pattern book. November 18, 2004.

4. N. GOKARNESHAN. Fabric structure and design. Copyright © 2004, New Age International (P) Ltd, Publishers Published by New Age International (P) Ltd, Publishers.

5. K. L. Gandhi. Woven textiles Principles, developments and applications© Woodhead Publishing Limited, 2012.

6. Prabir Kumar Banerjee. Principles of FABRIC FORMATION. © 2015 by Taylor & Francis Group, LLCCRC Press is an imprint of Taylor & Francis Group, an Informa business.

7. Martynova, Anna Arkhipovna. Structure and design of fabrics: Textbook for students of higher educational institutions / Martynova, Anna Arkhipovna, Slostina, Galina Leonidovna, Vlasova, Nina Aleksandrovna. - M. Izd-vo MSTU, 1999. - 434 8.Martynova A, A. Laboratory practice on structure and design of fabrics. Light Industry, 1976.

9. A R Horrocks and S Anand. HANDBOOK OF TECHNICAL TEXTILES Edited by The Bolton Institute, UK, 576 pages, 2000.

10. Dan J. McCreight, James B. Bradshaw.WEAVERS HANDBOOK OF TEXTILE CALCULATIONS. Everett E. Backe, and Michael S. Hill, Institute of Textile Technology, USA, 105 pages, 2000.

11. WEAVERS HANDBOOK OF TEXTILE CALCULATIONS. Dan J. McCreight, James B. Bradshaw, Everett E. Backe, and Michael S. Hill, Institute of Textile Technology, USA. Hill, Institute of Textile Technology, USA